Climate Change Resilience in the Urban Environment (Second Edition)

Online at: https://doi.org/10.1088/978-0-7503-5262-8

Climate Change Resilience in the Urban Environment (Second Edition)

Tristan Kershaw
Department of Architecture and Civil Engineering, University of Bath, Bath, UK

IOP Publishing, Bristol, UK

ISBN 978-0-7503-5262-8 (ebook)
ISBN 978-0-7503-5260-4 (print)
ISBN 978-0-7503-5263-5 (myPrint)
ISBN 978-0-7503-5261-1 (mobi)

DOI 10.1088/978-0-7503-5262-8

Multimedia content is available for this book from https://doi.org/10.1088/978-0-7503-5262-8

Version: 20241101

IOP ebooks

British Library Cataloguing-in-Publication Data: A catalogue record for this book is available from the British Library.

Published by IOP Publishing, wholly owned by The Institute of Physics, London

IOP Publishing, No.2 The Distillery, Glassfields, Avon Street, Bristol, BS2 0GR, UK

US Office: IOP Publishing, Inc., 190 North Independence Mall West, Suite 601, Philadelphia, PA 19106, USA

This book is dedicated to my family; my lovely wife Caroline and our children Jasper and Leonie, who excel at distracting me from writing and love to edit my work when I'm not looking.

Contents

Preface

For the first time in history more than half of the world's population lives in urban areas, a number that was only 30% in the 1960. This figure is set to increase dramatically over the next few decades, reaching ~70% by 2050. This rapid growth in urban populations will occur mainly in the developing world, with some countries expected to experience up to a five-fold increase in urban populations by 2050. Increasing urbanisation brings with it a series of benefits but also challenges. Increased urbanisation and urban density allow for more efficient use of materials and infrastructure, such as fewer roads, more efficient public transport, reduced distribution losses in energy networks and the ability to have larger more centralised public services (e.g., police, fire services and healthcare). However, there are drawbacks, increasing urban area typically means a loss of vegetative cover in favour of hard impermeable man-made surfaces. This can increase the risk of flooding from intense rainfall, while the dense materials used to construct our urban areas stores heat from the Sun to be radiated later, altering the local climate. Even the position and orientation of our buildings can have a negative effect, isolating the urban microclimate from the atmosphere above, trapping heat and pollutants at street level.

Human civilisation and current architectural practices have evolved over the past few thousand years during a period of relatively constant climate and predictable weather. As such, our buildings are the result of the interaction between culture, geography and meteorology. This has allowed architectural design and the layout of urban areas to be tailored to the local climate. Underneath the influences of culture and social interaction, the architecture of a building has two main functions: to provide shelter from the elements and to provide thermal comfort for its occupants. At every latitude, buildings transform sometimes harsh exterior environmental conditions into a comfortable interior environment. They achieve this feat through little more than careful design and the appropriate use of materials. However, as cities have begun to grow rapidly, these traits have been lost (figure P.1).

Globalisation has led to the prevalence of new materials and techniques, creating a divergence between the design and construction of buildings and their local cultural, social, ecological and environmental context. Buildings are now built more for speed of construction and cost rather than to be resilient to the local climate because cheap energy has meant that artificial heating and cooling can be used to compensate for a climate insensitive design.

Climate change is altering the way our buildings will need to perform. Changing temperatures, the seasonality of rainfall and increased instances of weather that are currently classed as extreme mean that our buildings will have to cope with different environmental conditions than previously. There is evidence, however, that vernacular architecture can be integrated into modern buildings to promote resilience to environmental stresses. But in order to combat the impacts of climate change, we will need to consider not only the vernacular architectural details from our own culture and climate but also those of other global locations which currently have a climate that may be similar to what we can expect in the future.

Figure P.1. Growing cities and expanding urban areas can provide many socioeconomic benefits but also provide many climate related challenges.

This book aims to consider not only what the impacts of climate change are on the built environment but also what we can do to try and mitigate against the impacts of climate change.

Acknowledgements

Research is often a collaboration with others, as such I would like to thank my students; Abdulla Alnuaimi, Chunde Liu, Kanchane Gunawardena, Petros Ampatzidis, and Yasser Ibrahim whose work appears in this book.

Author biography

Tristan Kershaw

Tristan Kershaw is an Associate Professor in Climate Resilience at the University of Bath. Tristan graduated from the University of Exeter in 2004 with a Master's degree in Physics and went on to study for a PhD in low temperature solid state physics. After completing his PhD, he joined the Centre for Energy and the Environment, also at the University of Exeter, as a research fellow in climate change adaptation. Over the subsequent six years, he worked on a variety of 'building physics' related research and consultancy projects, including the creation of probabilistic future weather years for the UK for the thermal modelling of buildings, and the modelling and the adaptation of building designs for several exemplar buildings across the southwest region. In 2014, Tristan joined the Department of Architecture and Civil Engineering at the University of Bath, teaching both undergraduate and postgraduate students, engineers and architects on the topics of building physics, sustainability, climate change and the dynamic modelling of building designs.

This book is based primarily upon my research and consultancy work, and as such contains a lot of case studies from in and around the southwest of the UK, along with observations of architecture and building use from further afield.

IOP Publishing

Climate Change Resilience in the Urban Environment
(Second Edition)

Tristan Kershaw

Chapter 1

Climate change and its impacts

Climate is a difficult concept for people to deal with because generally we think in terms of the short-term variations or weather and our memory is drawn towards more extreme events such as heat waves, cold snaps and storms. Climate however, is defined as the long-term averages and ranges of different weather variables. Changes in climate typically take many thousands of years; hence, human civilisation has evolved during a period of relatively constant climate. This means that buildings, urban areas and even human physiology are ill adapted to relatively rapid changes in climate over several decades or centuries. This chapter will examine the origins of climatic change, the evidence base and will form the basis for the subsequent chapters.

1.1 Introduction

The world is experiencing a massive transition from rural to urban living. Urban areas occupy less than 2% of the Earth's land surface but house over 50% of the world's population, a figure that was only 14% in 1900 and one which is estimated to increase to 60% by 2030. This rapid rise will mainly take place in developing countries. Asia and Africa will see the largest increases in urban population, which are expected to reach 64% and 56% urban population, respectively, by 2050. Countries such as Mali, Niger, Tanzania, Uganda and Zambia are predicted to see greater than a five-fold increase in urban population by 2050. If the developing world continues to urbanise as expected, replacing sprawling urban areas with extremely dense ones, they will face a plethora of problems. Denser urban areas are generally considered to be more sustainable, requiring less land, infrastructure and being resource efficient, and it is highly probably that future cites in currently developing countries will be at least as dense as those in the developed world. However, this comes at the cost of poorer air quality, reduced biodiversity, reduced

doi:10.1088/978-0-7503-5262-8ch1

flood resilience, higher air temperatures due to the urban heat island, and potentially poorer physical and mental health associated with reduced green (grass, trees and vegetation) and blue space (reservoirs, lakes, rivers, canals etc).

Climate change will affect the performance of buildings, urban areas, and the comfort and health of a city's occupants. This chapter will cover the origins and the scientific basis for climate change before presenting an overview of the impacts of climate change based upon the latest climate projections. Using these projections, the following chapters will examine the potential impacts of climate change on urban areas and consider how to increase the resilience of our growing urban areas to a rapidly changing climate.

1.2 The greenhouse effect

The climate the Earth experiences is a function of its distance from the Sun and the amount of radiation (energy) received. At any given time, the surface of the Earth must be in equilibrium with its surroundings, and therefore the Earth must be reradiating an equal amount of energy to that received from the Sun. All objects at a temperature above absolute zero (0 K, −273 °C) emit radiation. The hotter the object, the shorter the mean wavelength (λ) of this radiation. The peak wavelength of the distribution (black-body curve) of radiation emitted, λ_{max} (in metres, m), is given by Wien's displacement law:

$$\lambda_{max} = \frac{2.9 \times 10^{-3}}{T}$$

where, T (in Kelvin, K) is the temperature of the object. For the Sun, with a surface temperature of ~5800 K, the peak is at ~500 nm (500 nanometres or 500×10^{-9} m), which is in the visible light region. For the Earth, with its much lower temperature, the radiation is of longer wavelength and within the infrared part of the electromagnetic spectrum.

The Sun is not a perfect black-body emitter and not all the wavelengths of radiation given off reach the Earth's surface. Ozone (O_3) in the atmosphere absorbs and reradiates ultra-violet (UV) light. Rayleigh scattering by dust particles in the upper atmosphere, scatters blue light, hence why we view the sky as blue. Other wavelengths of light primarily in the near-infrared region are absorbed and reradiated by other gases in the atmosphere such as oxygen (O_2) and water vapour (H_2O). These absorption windows are associated with different mechanisms of bending and stretching of molecular bonds. The solar spectrum both at the top of the atmosphere and what reaches the surface of the Earth can be seen in figure 1.1.

Since we know the temperature of the Sun from its colour and spectral output, and we also know the distance of the Earth from the Sun, we can work out how much radiation the Earth receives, and hence the temperature the Earth should be to be in equilibrium, radiating an equivalent amount of energy to that it receives.

A hot object will radiate with an intensity I (W m^{-2}, Watts per square metre) given by:

$$I = \sigma T^4$$

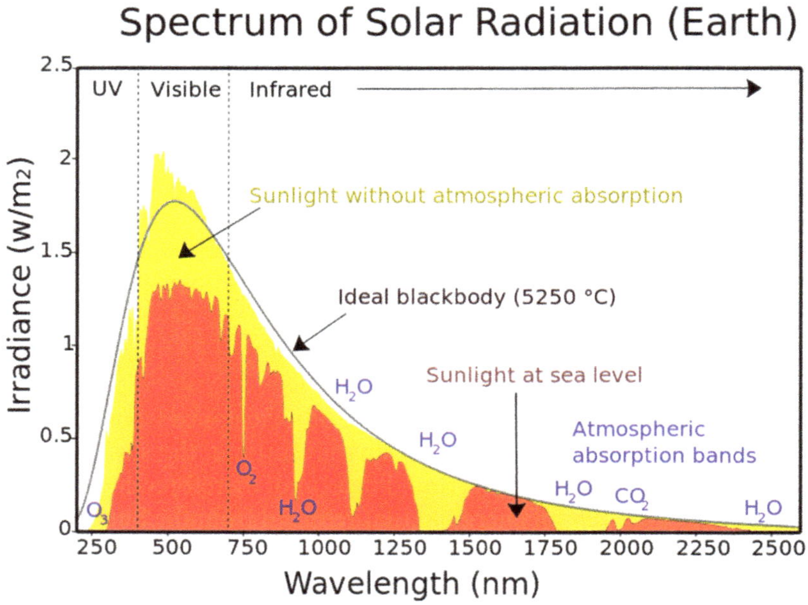

Figure 1.1. Solar spectrum at the top of the atmosphere and at sea level. This Solar spectrum image has been obtained by the author from the Wikimedia website where it was made available by under a CC BY-SA 3.0 licence. It is included within this article on that basis. It is attributed to Robert A. Rohde (User:Dragons flight).

where σ is the Stefan–Boltzmann constant (5.67×10^{-8} W m^{-2} K^{-4}). The surface of the Sun is at around ~5777 K, and hence radiates 6.32×10^7 W m^{-2}. The solar radius is 6.96×10^8 m (696 000 km), as such the surface area ($4\pi r^2$) of the Sun is 6.1×10^{18} m^2 and therefore must radiate with a luminosity of 3.85×10^{26} W. Assuming this radiation is emitted equally in all directions, we can estimate how much of this is intercepted by the Earth. If we imagine this radiation smeared across the inside of a sphere centred on the Sun and intersecting the Earth as shown in figure 1.2, then we can work out the fraction of the energy emitted intercepted by the Earth from the ratio of the two surface areas, the sphere and the Earth.

The Earth is at a distance of one Astronomical Unit (1 AU) from the Sun, which is equivalent to 1.496×10^{11} m (149.6 million km). Therefore, the surface area of the sphere intersecting the Earth is 2.81×10^{23} m^2. The Earth will project an area of πr^2 onto the surface of this sphere, which with a radius of 6.371×10^6 m (6371 km) equals 1.275×10^{14} m^2. Taking the ratio between the surface area of the sphere and the area the Earth projects on the surface of this sphere, we can calculate that the Earth receives only $1/4.5 \times 10^{10}$ th of the radiation from the Sun. Or, in simpler terms the Earth receives 1.74×10^{17} W from the Sun.

Not all the radiation received by the planet is absorbed. The Earth has an albedo of ~0.3, meaning that ~30% is reflected back directly into space from the atmosphere and the ground. This albedo varies slightly according to the amount of cloud in the atmosphere and where it is situated. Thus, the planet absorbs

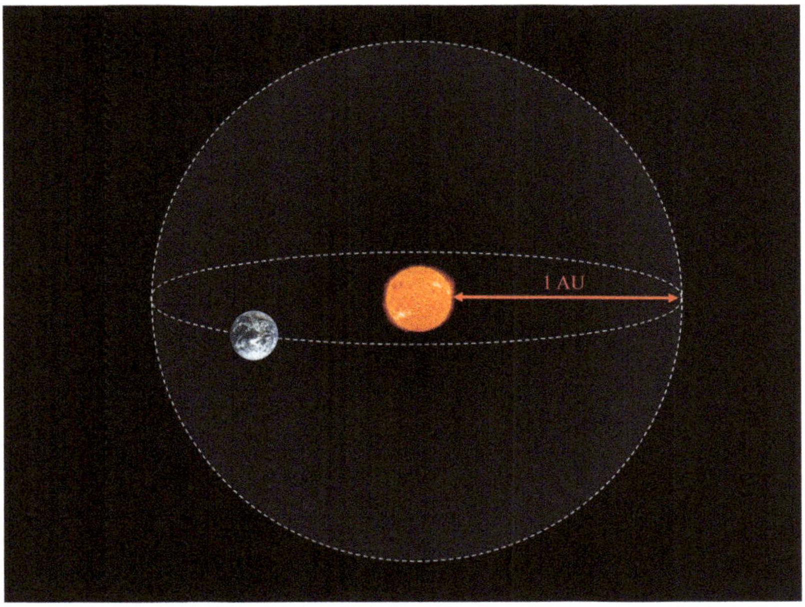

Figure 1.2. Consider a sphere of radius 1 AU centred on the Sun, with the Earth intersecting the surface.

approximately 1.22×10^{17} W, which when averaged over the whole surface of the Earth ($4\pi r^2$) is approximately $\sim$239 W m^{-2}. The planet must be in radiative balance (equilibrium), reradiating all radiation received; else the planet's temperature would continuously escalate. Hence, working backwards using $I = \sigma T^4$, we can estimate the temperature of the Earth required to be in radiative balance. This suggests that the temperature of the Earth should be $\sim$255 K, or –18 °C. While this value calculated is close to the measured temperature of the Moon, clearly this is not true for the Earth. The average temperature of the Earth is closer to 15 °C (298 K), the reason for this, which all scientists agree on, lies in the makeup and chemistry of the Earth's atmosphere.

This atmospheric chemistry, which is responsible for the $\sim$33 °C disparity between the observed and calculated temperature of the Earth's surface, is termed the natural greenhouse effect. The basis of this effect lies in the relative absorption spectra of the different atmospheric gases. Figure 1.3 shows the Earth in the visible spectrum, the result is the iconic blue marble that we are all familiar with, indicating that the entire visible spectrum is able to pass unhindered through the atmosphere to the surface. Light from the Sun travels through space, through the atmosphere and hits the surface of the Earth, in the case of figure 1.3 a grain of sand in the Sahara, is reflected and travels back through the atmosphere of the Earth to the camera lens, unobstructed and without scattering. The fact that we can see the surface of the Earth from space in great detail tells us that we have not lost any directional information, light is not scattered, diverted or absorbed during its passage through the atmosphere. The only scattering comes from clouds in the atmosphere; appearing white, we can deduce that they scatter all visible light equally. The result

Figure 1.3. The Earth viewed from space in the visible light spectrum. Image from AFLO/naturepl.com.

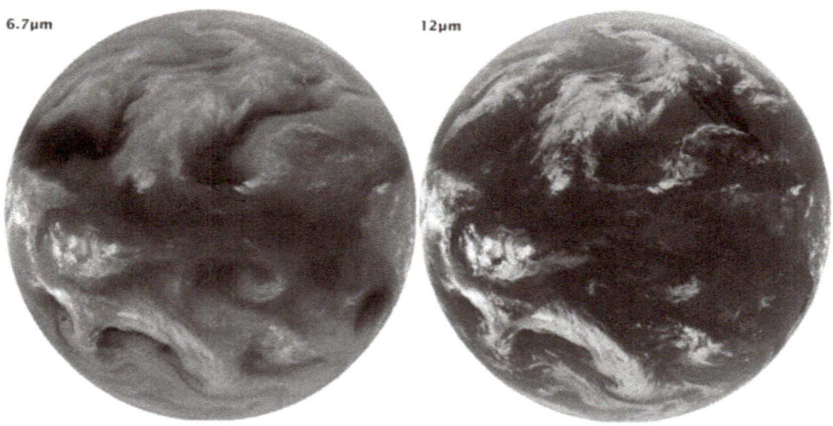

Figure 1.4. The Earth viewed from space in the infrared spectrum. Courtesy of NASA.

is that of the solar radiation intercepting the Earth, around 70% is absorbed and contributes to heat the surface, while ~30% is reflected from the surface or from clouds in the atmosphere, the atmosphere itself is transparent. For comparison, figure 1.4 shows the Earth viewed at two wavelengths in the infrared spectrum. The left-hand image shows the Earth viewed at 6.7 μm (6.7 microns or 6.7×10^{-6} m), which is within a water vapour absorption band. We can see that we have lost all directional information, light is being fully scattered (absorbed and reradiated) within the atmosphere. The right-hand image shows the same image of the Earth but viewed at 12 μm, which while in the infrared spectrum is largely transmitted by the atmospheric greenhouse gases. Here, we can see that we have lost some directional information, the image is milky not crystal clear but we can still make out the continents. This is the origin of the natural greenhouse effect and the additional warming the Earth receives as a result. Visible light is able to pass relatively unhindered through the atmosphere to the surface of the planet, where it warms the surface. The warm surface of the planet radiates according to its temperature in the infrared spectrum. However, these wavelengths of light find it much harder to pass through the atmosphere, being reflected back to the surface, absorbed and reradiated by gases in the atmosphere. The absorption (and transmission) spectra of the atmosphere as a whole and the most important greenhouse gases are shown in figure 1.5.

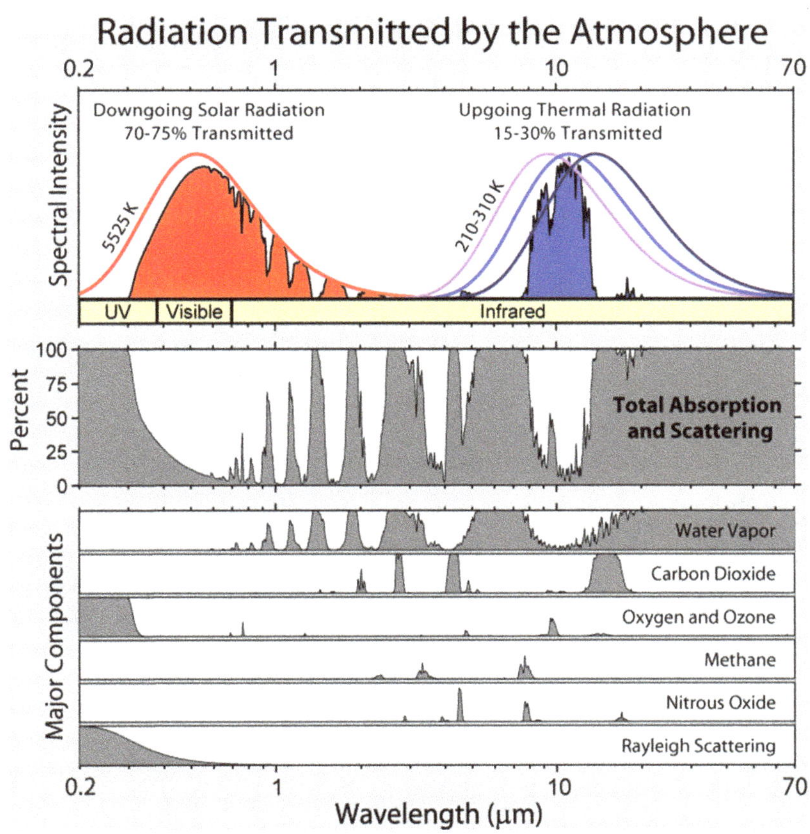

Figure 1.5. Plot of the radiation transmitted through the atmosphere (top) and the relative absorption spectra of the various atmospheric greenhouse gases (bottom). This Atmospheric Transmission image has been obtained by the author from the Wikimedia website where it was made available by under a CC BY-SA 3.0 licence. It is included within this article on that basis. It is attributed to Robert A. Rohde (User:Dragons flight). https://commons.wikimedia.org/wiki/File:Atmospheric_Transmission.png

The absorption spectrum arises from the absorption of light and its conversion into mechanical energy, in this case vibrations of molecular bonds and the motion of the constituent atoms. This energy is transferred to the surrounding gases (principally nitrogen and oxygen, which are not greenhouse gases), leading to a warming of the atmosphere.

The result being that although we have a radiative balance at the top of the atmosphere for the planet to be in equilibrium, for the surface and the rest of the atmosphere we need a more complex approach. We need to take account of the absorption, reflection, heat transfer and convection within the atmosphere. As we can see from figure 1.5, different atmospheric gases absorb different wavelengths of light, these gases must reradiate all the energy they absorb, but unlike the surface of the Earth they are able to radiate in all directions, not just upwards. Hence, some of this radiation will be reradiated back towards the Earth's surface and contribute to additional heating. It is this natural greenhouse effect that produces the elevated surface temperatures and results in the average temperature of the Earth being 15 °C rather than –18 °C. We can see from figure 1.5 that water vapour is the largest contributor to atmospheric absorption and of the 33 °C additional warming experienced as a result of the natural greenhouse effect, water vapour is responsible for approximately 22 °C of it.

Figure 1.6 provides an illustration of the typical magnitudes of the radiative flux experienced through the atmosphere. Of the 341 $W\,m^{-2}$ that hits the top of the atmosphere, around 30% is reflected back out to space (net downwards radiation absorbed 239 $W\,m^{-2}$). This value must be equivalent to the total outgoing radiation

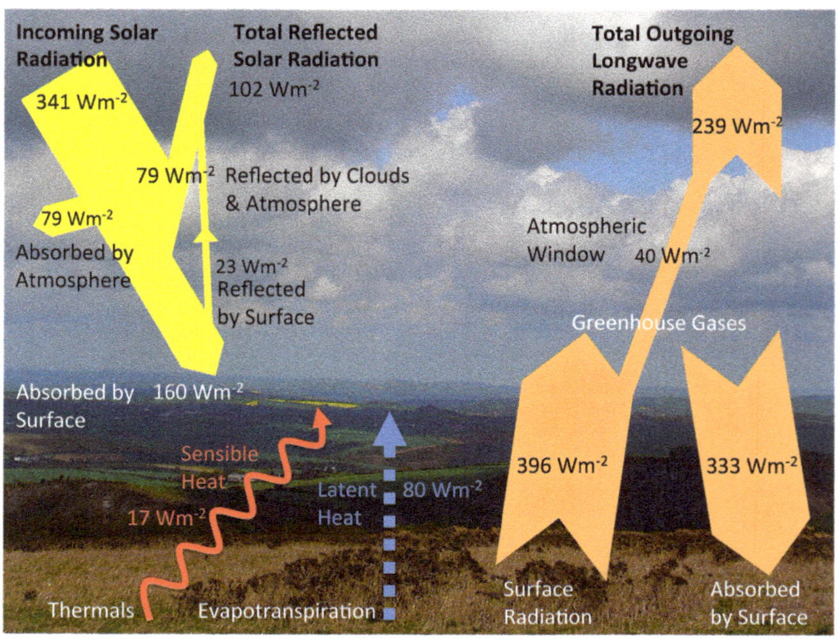

Figure 1.6. Representation of the average radiation balance of the atmosphere. These values will vary according to season, latitude and with astronomical mechanics.

for the Earth to be in equilibrium. The atmosphere absorbs 79 W m^{-2}, while 160 W m^{-2} is absorbed by the surface. Around 80 W m^{-2} goes into the evaporation of water and transpiration of water by plants (collectively referred to as evapotranspiration) and drives the hydrological cycle, while around 17 W m^{-2} is used to set up convection currents within the atmosphere. Ultimately, the total up going radiation from the Earth's surface is 396 W m^{-2}, of which 333 W m^{-2} is reflected or radiated back towards the surface by the atmosphere. It is this that produces the higher than expected average surface temperatures. There is only a small window of wavelengths where heat (infrared radiation) can pass through the atmosphere unimpeded, accounting for only 40 W m^{-2}, the remainder of this radiation goes into warming of the atmosphere, as do ultimately the sensible and latent heat contributions.

1.3 The historic climate signal

While human civilisation has arisen during a period of relatively constant climate, the Earth has seen many different climates over its 4.5 billion year history. Through examination of the fossil record and deep ice cores, we can draw a picture of what life on Earth was like at different time periods.

There are stable isotopes of oxygen, oxygen-16 (^{16}O), which contains eight protons and eight neutrons, and the less common oxygen-18 (^{18}O), which contains eight protons and 10 neutrons. In the paleosciences, the ratio of ^{18}O : ^{16}O ($\delta^{18}O$) found in corals, fossils and ice cores can be used as a proxy for temperature. This arises from the differential rates at which water molecules containing these isotopes evaporate or condense. When water vapour condenses, the heavier water molecules containing ^{18}O atoms condense and precipitate first. As such, there is a preferential evaporation of ^{16}O from seawater, hence fresh water precipitation is ^{16}O enriched, leading to a gradient in the $\delta^{18}O$ with latitude. The surface of the oceans contains greater amounts of ^{18}O around the tropics where there is increased evaporation and reduced amounts of ^{18}O at the mid-latitudes where there is more rain. Additionally, the amount of ^{18}O present in water vapour is greater at the tropics than closer to the poles due to higher temperatures and greater evaporation. Snow that falls in Russia or Canada has much less $H_2{}^{18}O$ than rain that falls in Malaysia or Peru. Similarly, snow falling at the centre of an ice sheet will have less ^{18}O, than snow falling at the edges of the ice sheet, due to the preferential condensation of ^{18}O and $H_2{}^{18}O$ precipitating first. From the $\delta^{18}O$ ratio, we can infer the temperature of precipitation, and hence how much warmer or colder the Earth was at the time the snow fell.

Additionally, study of Antarctic ice cores examining the ratio of oxygen to nitrogen in bubbles within the ice can be used to infer the level of insolation (solar radiation intensity). These bubbles within the ice can also be analysed to determine the concentrations of greenhouse gases at the time such as carbon dioxide (CO_2) and methane (CH_4). Resultant changes in climate from differing greenhouse gas concentrations alter the patterns of global evaporation and precipitation, and therefore change the ratio $\delta^{18}O$. Data from Vostok Station in Antarctica (shown in figure 1.7) shows that the Earth's climate has varied considerably over previous millennia, with peaks and troughs in temperature.

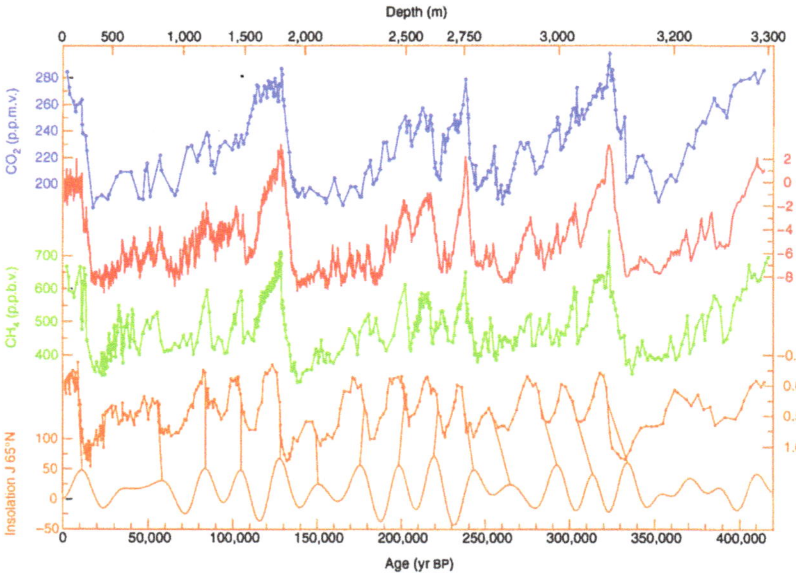

Figure 1.7. Ice core data from Vostok Station in Antarctica, showing $\delta^{18}O$ and associated temperature estimation. Also shown are insolation, CH_4 (parts per billion by volume) and CO_2 (parts per million by volume) concentration from gas bubbles within the ice cores. This Vostok 420ky 4curves insolation image has been obtained by the authors from the Wikimedia website, where it is stated to have been released into the public domain. It is included within this article on that basis. https://commons.wikimedia.org/wiki/File:Vostok_420ky_4curves_insolation.jpg

Importantly, we can see the temperature profile closely follows the concentrations of CO_2 and CH_4. Furthermore, the temperature profile follows some periodicity related to the level of insolation. Greater insolation means greater evaporation of water, and hence more water vapour in the atmosphere (the most important greenhouse gas). But why does the level of insolation vary with time and why is there periodicity?

Unlike early pictures of the Solar System, the planets do not hold circular orbits around the Sun. The Earth spins on its axis and orbits the Sun; however, there are several quasi-periodic interactions due to the gravitational forces from other celestial bodies. The Serbian mathematician Milutin Milanković (1879–1958) studied the changes in the Earth's orbital eccentricity, obliquity and precession, and their influence of the level of insolation. These observed cycles in insolation, and hence cycles in the Earth's temperature, are now generally referred to as Milankovitch Cycles.

Unlike our idealised example earlier in this chapter, the Earth's orbit is not circular and the distance from the Sun to the Earth is not a constant. The Earth's orbit is an ellipse. Eccentricity is a measure of the departure of an ellipse from being circular (see figure 1.8), the eccentricity of a circle is by definition 0, while a parabola has an eccentricity of 1 and an ellipse has an eccentricity <1. The equation for a circle is of the form

$$x^2 + y^2 = r^2$$

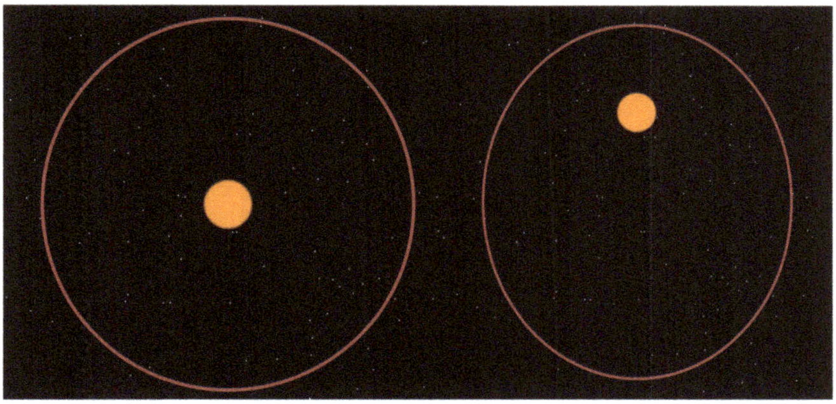

Figure 1.8. Illustration of a circular orbit (left-hand) and an elliptical orbit with highly exaggerated eccentricity (right-hand). Courtesy of NASA.

while an ellipse has the form

$$\frac{x^2}{a^2} + \frac{y^2}{b^2} = 1$$

and has an eccentricity (e) of the form

$$\sqrt{1 - \frac{b^2}{a^2}}$$

The shape of the Earth's orbit varies with time between nearly circular, with a low eccentricity ($e = 0.000\,055$) and a more elliptical orbit ($e = 0.0679$). The mean eccentricity of the Earth's orbit is considered to be $e = 0.0019$. Figure 1.8 provides an illustration of both a circular and a highly eccentric elliptical orbit.

If the Earth was the only planet orbiting the Sun, then the eccentricity of the orbit would not perceptible alter over millions of years. However, the presence of other large planets in our Solar System, such as Saturn and Jupiter, produces cyclic interactions due their gravitational field, which vary with their orbit around the Sun and the Earth's position relative to them. As you might expect with seven other planets and dwarf planets (including Pluto, Ceres and Eris), predicting variations in eccentricity due to so many gravitational interactions is complex. However, the signal is dominated by the largest bodies, resulting in a few main components for the periodicity of the eccentricity. The main component has a period of 413 000 years ($e \pm 0.012$). While other components have periods of 95 000 years and 125 000 years, resulting in a beat frequency of ~400 000 years. The combinations of these signals loosely combine into a 96 000 year cycle (see figure 1.12) with a magnitude between –0.03 and +0.02. Presently, the Earth's orbit has an eccentricity of $e = 0.017$ and is decreasing.

As eccentricity increases, variations in seasons increase due to varying distance from the Sun and varying insolation. However, this variation in insolation due to eccentricity is generally small, the intensity of the seasons is not primarily governed by the distance from the Sun, but rather the axial tilt of the Earth (figure 1.9). The

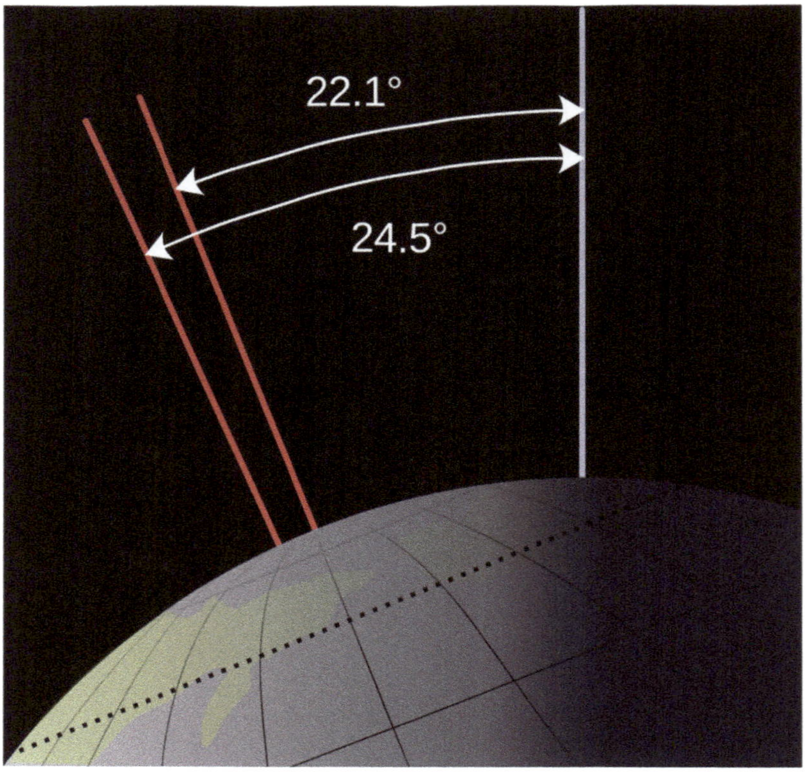

Figure 1.9. Illustration of changes to the Earth's axial tilt or obliquity (orbital motion is out of the page). Courtesy of NASA.

relative increase in insolation at the perihelion (the closest approach to the Sun) compared to insolation at the aphelion (furthest distance from the Sun) is roughly four times the eccentricity. Currently, this amounts to a variation in insolation of ~6.8%. Perihelion currently occurs around the 3rd of January, while aphelion is around the 4th of July.

Changes to the Earth's eccentricity do not change the length of a year or alter its motion along its orbit. However, higher eccentricity does exaggerate behaviour due to precession and axial tilt, which are explained below.

As we all know, the Earth's axis of rotation is not upright (perpendicular to the direction of orbital motion), instead the Earth spins around an axis that is ~23° off perpendicular. It is this axial tilt (obliquity) that is responsible for the seasons we experience, and the long polar nights and days. The angle of the Earth's axial tilt varies over time with respect to the planets orbital plane. Axial tilt varies by up to 2.4° between 22.1° and 24.5° in a periodic way, taking approximately 41 000 years. When the obliquity increases, the magnitude of the seasonal variations in insolation increase. Summers receive more radiative flux and winters comparatively less. The opposite is true as obliquity decreases. It should be noted that these changes to the magnitude of insolation are not equal over the entire Earth's surface. At higher latitudes the mean annual insolation increases with obliquity, while at lower latitudes obliquity decreases insolation. As such, it can be argued that lower obliquity favours the onset of ice ages by reducing both the overall mean summer insolation and additionally reducing insolation at higher latitudes. This results in reduced melting of the previous winters precipitation because most of the Earth's snow and ice lies at higher latitudes. The resulting change in the Earth's albedo can result in further cooling, creating an ice age, which will persist until the obliquity increases.

Currently, the obliquity of the Earth is 23.44°, which is about halfway through its cycle and is decreasing. The obliquity will reach its minimum around 11 800 AD and was last at its maximum in 8700 BC. Therefore, we can say that the Earth is currently experiencing a cooling trend.

The Earth not only spins about a tilted axis, which varies with time, but also 'wobbles' like a spinning top. This process, termed precession, is the trend in the direction of the Earth's axis of rotation to rotate relative to the orbital plane and the fixed stars (see figure 1.10). This axial precession has a period of roughly 26 000 years

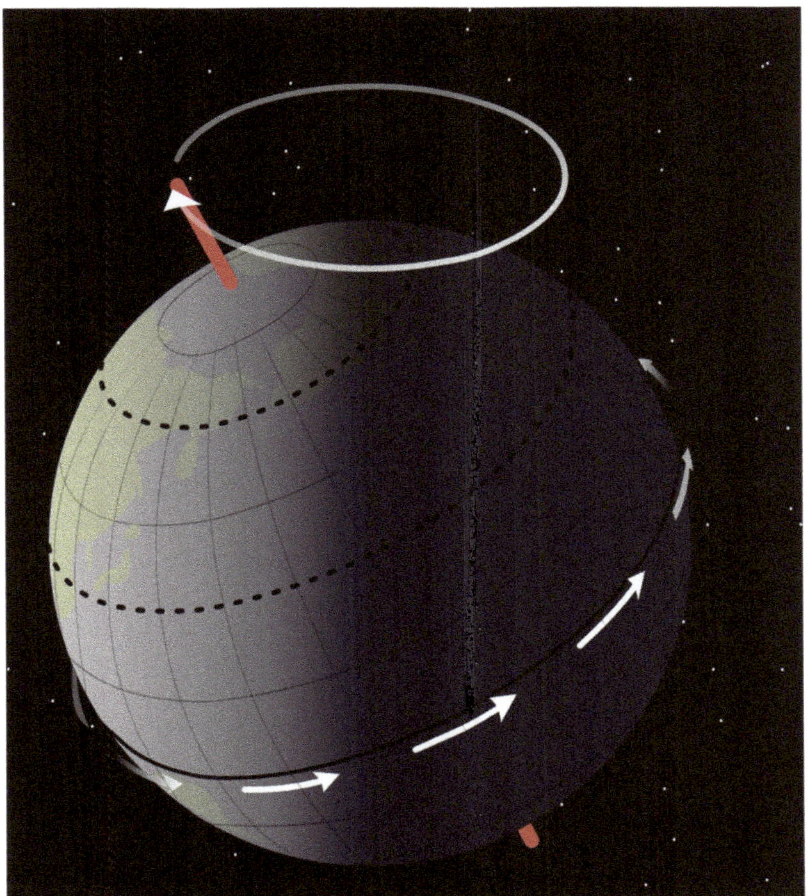

Figure 1.10. Illustration of the Earth's axial precession. Courtesy of NASA. Video available at http://iopscience.iop.org/book/mono/978-0-7503-5262-8.

and is due to gravitational tidal forces exerted by the Sun and the Moon upon the Earth. The contribution from each is roughly equal.

When the Earth's axis of rotation points towards the Sun at perihelion (i.e. North pole is pointed toward the Sun), the Northern hemisphere has a greater variation in seasons while the Southern hemisphere experiences reduced seasonal variability (milder seasons). While the inverse is true when the axis of rotation points away from the Sun at perihelion (i.e. South pole points towards the Sun), resulting in greater seasonal variations in the Southern hemisphere and milder seasons in the Northern hemisphere. The hemisphere that is summer perihelion will receive most of the increase in insolation resulting in higher summertime temperatures, but correspondingly will experience much colder winters when it is in aphelion. At present, the Southern hemisphere is in perihelion during the summer and aphelion during the winter, and as such experience the greater variation in seasonal extremes relative to the Northern hemisphere, which is experiencing milder seasons.

In addition to eccentricity, obliquity and axial precession, the Earth's elliptical orbit precesses in space (see figure 1.11). This orbital (apsidal) precession completes

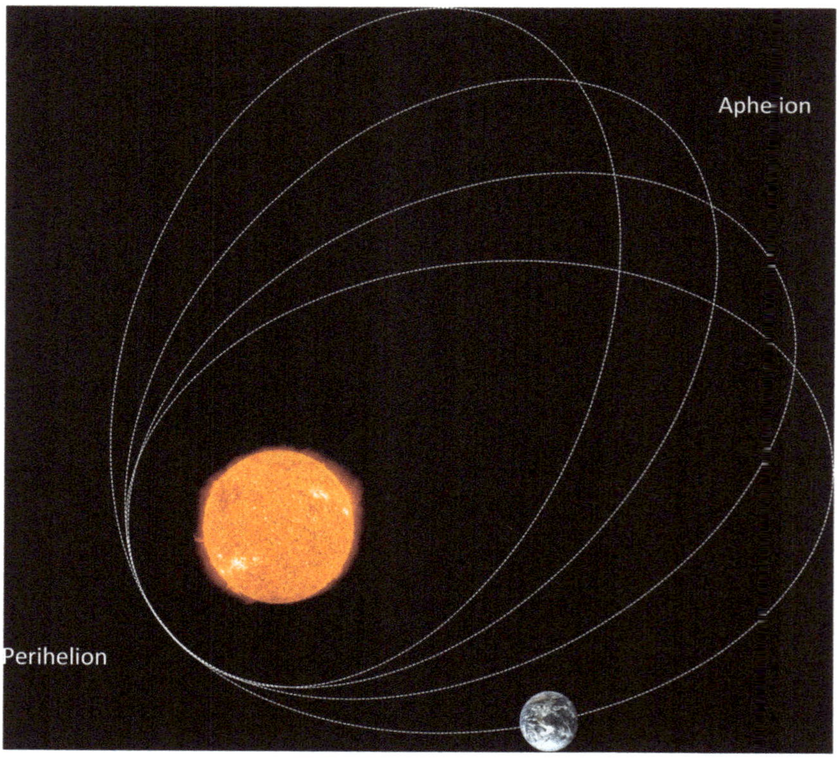

Figure 1.11. Illustration of apsidal precession for an exaggerated highly eccentric elliptical orbit. Video available at http://iopscience.iop.org/book/mono/978-0-7503-5262-8.

a cycle every 112 000 years relative to the fixed stars. Apsidal precession occurs in the elliptic plane and alters the Earth's orbit relative to the elliptic. In combination with changes in eccentricity, this alters the length of seasons.

All of these changes in the movement of the Earth relative to the Sun alter the amount of insolation the Earth receives. The sum of all these different cycles in insolation produces a complex signal of temperature oscillation, which can be observed in both the fossil record and in ice cores. Figure 1.12 shows the cyclic contributions of the different orbital movements (Milankovitch cycles), the resultant changes in insolation and the observed variation in $\delta^{18}O$ from ice cores and within benthic foraminifera and the inferred variation in temperature. Benthic Foraminifera (forams for short) are single celled organisms related to amoeba, which produce a hard shell. Foram shells are made of calcium carbonate ($CaCO_3$) and are found in many common geological environments. The ratio $\delta^{18}O$ in the shell is used to determine the temperature of the surrounding water at the time the organism was alive. This can be used as another indicator along with ice core $\delta^{18}O$

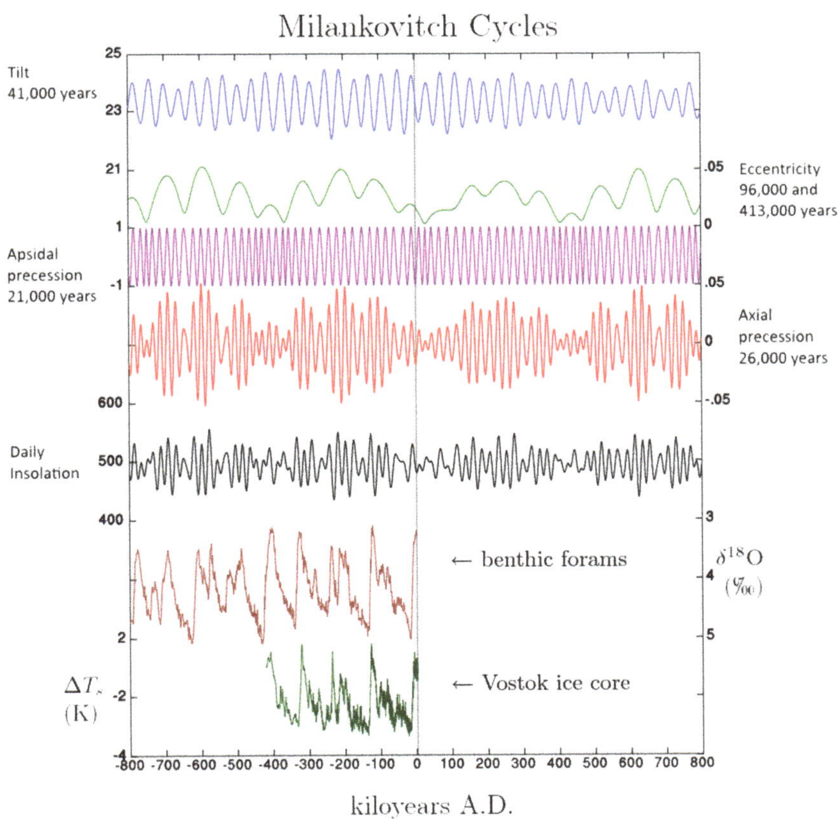

Figure 1.12. Plot of the various components of Milankovitch cycles and the observed $\delta^{18}O$ and temperature from ice cores and the fossil record. This MilankovitchCyclesOrbitandCores image has been obtained by the author from the Wikimedia website where it was made available by under a CC BY-SA 3.0 licence. It is included within this article on that basis. It is attributed to User:Incredio. https://commons.wikimedia.org/wiki/File:MilankovitchCyclesOrbitandCoresRecaptioned.png

and gas bubbles within the ice to determine what conditions were like on Earth in the past.

1.4 The anthropogenic greenhouse effect

As we have explored in the previous section, the make-up of the Earth's atmosphere is responsible for the climatic conditions we experience and for life as we know it. Given the vast volume of the atmosphere, it seems unlikely that human activities could influence the composition of the atmosphere and alter the energetic balance of the planet. The natural greenhouse effect was first described in 1859 by the British scientist John Tyndall, when he discovered that the most common components of the atmosphere—nitrogen and oxygen—were transparent to both visible and infrared radiation, whereas gases such as carbon dioxide, methane and water vapour were not transparent in the infrared. He concluded that such gases must have a great influence on our climate. In 1894, the Swedish chemist Svante Arrhenius showed that anthropogenic (man-made) emissions had the potential to alter the climate by further reducing the transparency of the atmosphere in the infrared spectrum. He further concluded that at the current rate of emissions it would take mankind 3000 years of burning coal to double the concentration of CO_2 in the atmosphere; in this last point, he was off by around 28 centuries!

While we can examine past climate through ice cores and forams, the resolution both temporally and spatially is quite low. Only relatively recently have we started to monitor the climate of the Earth in greater detail. The oldest running continuous series of temperature observations in the world is the Central England Temperature Record. Daily and monthly temperatures from three observations stations are used to produce representative measurements of a triangular area enclosing Lancashire, London and Bristol. Monthly measurements begin in 1659 and daily measurements begin in 1772. Figure 1.13 shows a plot of the mean annual temperature from 1659 to the end of 2023, created by averaging the monthly means.

While there is a large amount of variability in the temperature record, we can see that there are cooler and warmer periods. For example, we can make out the 'mini ice age' of the later seventeenth century as well as particularly cold or warm individual years. However, since the Industrial Revolution, we can see a steady rise in the temperature signal, despite the Earth being within a cooling cycle determined by orbital mechanics discussed in the previous section. This can be attributed to the changing concentrations of greenhouse gases within the atmosphere as a result of human activities, enhancing the already present natural greenhouse effect. Figure 1.14 shows the trend in atmospheric CO_2 since 1958, the so-called Keeling curve, based on the work started by Charles Keeling. Monthly measurements of atmospheric CO_2 concentration at the Mauna Loa observatory (Hawaii) by the National Oceanic and Atmospheric Administration (NOAA) show an accelerating trend in CO_2 concentration. Recently, atmospheric concentration passed 400 ppm (parts per million) for the first time since modern humans have walked the Earth (compare figure 1.14 to the concentrations shown in figure 1.7 from ice cores).

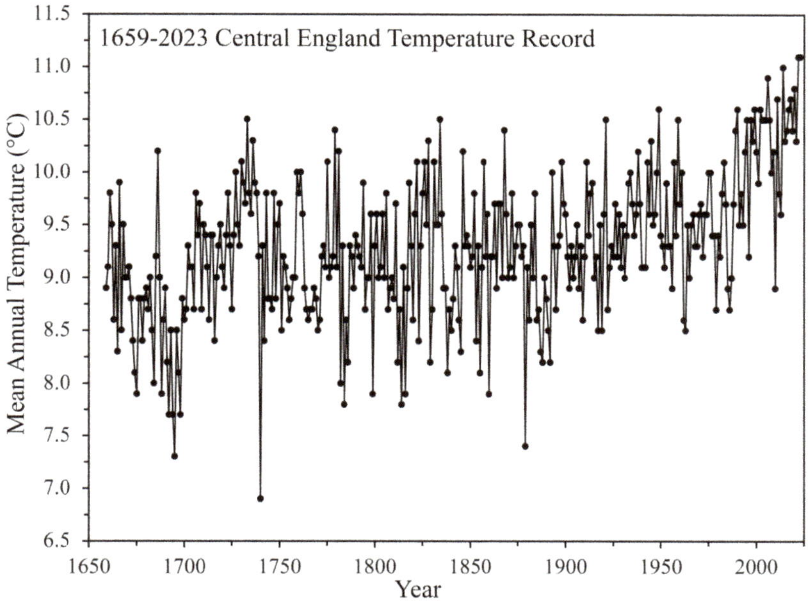

Figure 1.13. Plot of mean annual temperature from the Central England Temperature record[1].

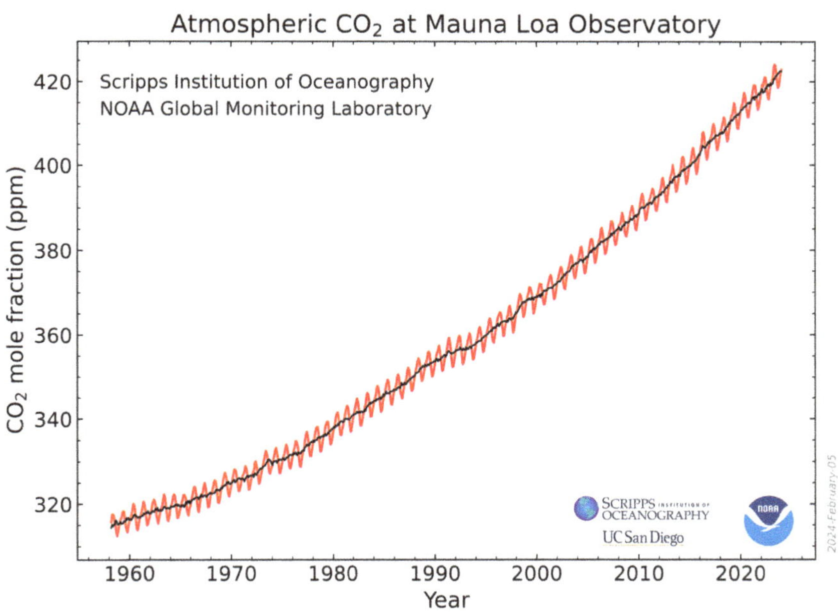

Figure 1.14. Plot of monthly measurements of CO_2 from Mauna Loa observatory, the red line represents monthly measurements while the black line is an annual mean. Reprinted with permission from NOAA https://gml.noaa.gov.

[1] Data: www.metoffice.gov.uk/hadobs under Open Government Licence v3.0 Citation: Parker D E, Legg T P, Folland C K 1992 A new daily Central England Temperature Series, 1772–1991 *Int. J. Clim.* **12** 317–42.

The oscillations in the atmospheric CO_2 (red line) are a result of the seasonal variations in the northern hemisphere. Since the majority of the Earth's land mass and forests are located in the northern hemisphere, the atmospheric CO_2 concentration is dominated by the northern summer and winter, due to the annual cycles of plant life. Worryingly, this trend in CO_2 concentration does not show any visible deviations due to the Rio Earth summit (1992), the Kyoto protocol (1997), Rio +20 (2012) or Paris Convention (2015), the only steps in the data are attributed to the collapse of the Soviet Union in the early-1990s, and the effects of the Covid 19 pandemic (2020–21).

There is often much confusion about the origin of this additional CO_2, whether it comes from volcanoes, deforestation or from the burning of fossil fuels. There is evidence, however, to show that the carbon emissions increasing the CO_2 concentration of the atmosphere are as a result of anthropogenic emissions, primarily from the burning of fossil fuel. Like Oxygen, Carbon exhibits several different isotopes, with different masses. Carbon in the atmosphere is ~99.89% Carbon-12 (^{12}C) and ~1.11% Carbon-13 (^{13}C), and trace amounts of Carbon-14 (^{14}C). ^{12}C and ^{13}C are stable but ^{14}C is radioactive with a half-life of 5730-years (half-life is the amount of time it takes for the radioactivity of a substance to halve as it decays). The length of this half-life means that any ^{14}C that was created when the Earth formed would long since have disappeared, implying that new ^{14}C must be constantly being made. This process occurs through highly energetic particles called cosmic rays entering the atmosphere, resulting in the production of free neutrons. These neutrons collide with Nitrogen atoms in the atmosphere producing ^{14}C and a proton.

$$^{1}n + ^{14}N \rightarrow ^{14}C + ^{1}p$$

This process occurs in the troposphere and stratosphere at an altitude of 9–15 km. The ^{14}C combines with oxygen molecules to from radioactive $^{14}CO_2$ at a concentration of one part in a trillion (1:10^{12}), which is taken up by plants and algae, passing up the food chain to other living organisms. This concentration is maintained while the organism is living, only reducing once the organism dies. This means that the amount of ^{14}C present can be used to date objects such as wooden axe handles or animal hides used for shelter by early humans or any carbonaceous material up to about 60 000 years old. By carbon dating the atmosphere, we can determine the source of the additional CO_2. If the increasing CO_2 concentration were the result of deforestation, then the amount of ^{14}C would remain constant, whereas we observe a decreasing concentration of ^{14}C in the atmosphere. This shows that the carbon emissions are coming from an old source of carbon such as the burning of fossil fuels which are old enough that all ^{14}C in the material has long since decayed. We can track the concentration of ^{14}C in the atmosphere over recent times through observation of tree rings and find that the decreasing concentration of ^{14}C coincides with the Industrial Revolution and has decreased in-line with increasing fossil fuel use. While volcanoes do produce substantial amounts of CO_2 and much of this would be old carbon with reduced ^{14}C, these events are not regular enough to explain the observed trends. Additionally, the eruption of volcanoes needs to be weighed against the disruption they cause to normal human activities. The recent eruption of the Icelandic volcano Eyjafjallajökull in 2010, could be considered a carbon negative event due to the widespread disruption to air travel across Europe. Hence, we can conclude that the observed trends in atmospheric CO_2 concentration are primarily as a result of human activities. Much of the confusion is likely due to the inertia of the carbon cycle and the atmosphere. It takes a long time for changes to the atmosphere to exhibit themselves as warming at the surface of the planet because both the land and oceans have high thermal inertia. We are only just starting to observe oceanic warming despite having been pumping out increasing amounts of CO_2 for over two centuries. Figure 1.15 overlays the observed CO_2 concentrations shown in figure 1.14 on top of the observed global temperature anomaly (variation). The baseline (0 °C) for this temperature anomaly is the twentieth century (1901–2000). Through close examination of the trends, we can see the lag between increasing CO_2 concentrations and increasing temperatures.

This thermal inertia also has a downside in that even if we were to stop all carbon emissions tomorrow, the land surface and oceans would continue to warm and sea levels would still continue to rise, until equilibrium is reached (see section 1.2). Of course, CO_2 is not the only greenhouse gas, water vapour is the most important in terms of the elevated temperatures the Earth experiences, but water vapour is not directly produced by human activities. There are, however, several other gases that have a significant greenhouse effect, notably methane, which is produced by extensive farming, particularly of cattle or rice. There are also several man-made chemicals that are very effective at trapping heat in the atmosphere, such as Chloroflurocarbons (CFCs), Sulphurhexafuoride (SF_6) and Hydroflurocarbons (HFCs). Not only do these different chemicals have different effectiveness at

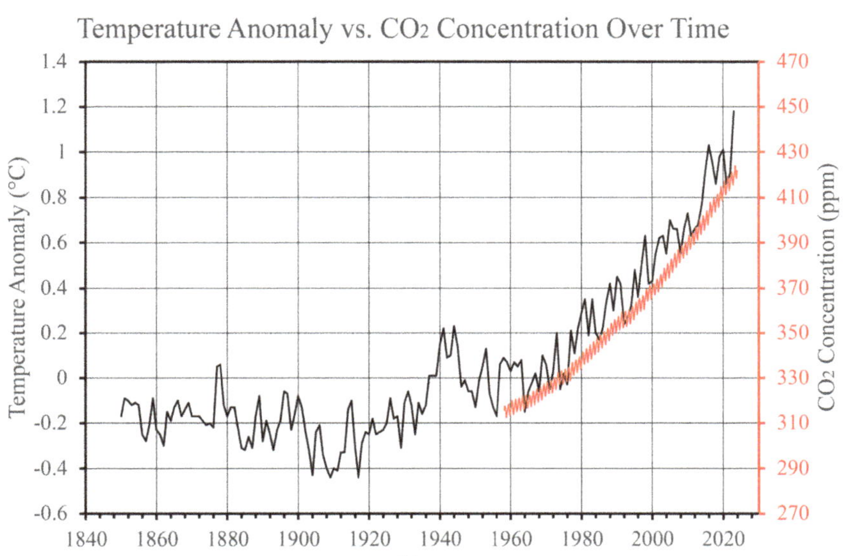

Figure 1.15. Plot of average land and ocean temperature anomaly and CO_2 concentration against time (data obtained from NOAA https://gml.noaa.gov).

Table 1.1. Lifetime and relative global warming potential of different atmospheric chemicals against a 100-year baseline of CO_2 accounting for climate-carbon feedback. (Source IPCC AR5 and AR4 for SF_6).

Name	Formula	Lifetime	Relative GWP
Carbon dioxide	CO_2	100	1
Methane	CH_4	12.4	34
Nitrous oxide	N_2O	121	298
Hydroflurocarbon	HFC-134a	13.4	1550
Chloroflurocarbon	CFC-11	45	5380
Carbon tetrafluoride	CF_4	50 000	7350
Sulphur hexafluoride	SF_6	3200	22 800

absorbing heat radiated from the Earth's surface but they remain in the atmosphere for different amounts of time. A CO_2 molecule typically remains in the atmosphere for between five and two-hundred years (generally accepted as one-hundred years) before it is absorbed by a plant and starts to move around the carbon cycle. Methane is only present in the atmosphere for around 12-years, but it is much more effective at trapping heat. As such, we need a measure to compare the relative effectiveness of these different chemicals in order to inform mitigation policies. Relative global warming potential (GWP) is used to make this comparison; CO_2 is used as a baseline and has a GWP = 1 for a 100-year horizon. Table 1.1 compares the different GWPs of several important anthropogenic greenhouse gases.

This also allows the equation of different greenhouse gas emissions as Carbon Dioxide equivalent (CO_2e or CO_2-eq). Increasing the atmospheric concentration of any of these gases will have a warming effect because more heat radiated from the Earth's surface will be intercepted and reradiated back at the surface. The term radiative forcing is used to quantify the additional amount of warming experienced. A doubling of CO_2 above pre-industrial levels (278 ppm) is expected to have a radiative forcing of $\sim$4 W m^{-2}, comparing this to the values shown in figure 1.6, we can see that this will have a profound effect on the climate we experience.

1.5 Climate change projections

The Intergovernmental Panel on Climate Change (IPCC) assessment reports summarise global climate change impacts. We will discuss projections of climatic change in greater detail in the following chapters of this book. Here, we will briefly summarise some of the projections from the IPCC and their origins.

For the first four IPCC assessment reports (1990–2007), estimates of future climate change were based upon socioeconomic scenarios, detailed in the IPCC 'Special Report: Emission Scenarios' or SRES for short. Future greenhouse gas emissions are a product of the complex interactions of many different dynamic systems. These SRES scenarios cover a wide range of driving forces that influence current and future emissions, including demographic, technological and economic developments. These scenarios include the range of emissions for all the relevant greenhouse gases and their driving forces. As such, the IPPC state that the likelihood of any single emissions path actually occurring is very small. As such, when dealing with future climate change, we are considering possible projections of what the world could be like, rather than definitive predictions of what the world will be like in 20, 40 or 100 years.

By the end of this century, the world will have changed in ways that are very difficult to imagine, in the same way those who lived at the turn of the last century would find it hard to predict today's way of life. The SRES scenarios consider four different storylines to describe consistently the interrelationships between emission driving forces and their evolution. Each storyline (A1, A2, B1 and B2) represents different demographic, social, economic technological and environmental developments. These storylines become increasingly divergent, in irreversible ways, and because of their complexity their feasibility cannot simply be based upon extrapolation of current socioeconomic trends. No probability or likelihood is associated with any of these storylines or scenarios. Some storylines such as B2 are becoming increasingly infeasible due to recent population growth, but other factors within the storyline are still possible. Hence, it cannot simply be ignored. The primary issue with the SRES storylines is their age. Created in 1990, they do not account for the rise of countries such as China and India as massive economic powers and the associated demographic, social and local technological improvements implemented, or the resultant carbon emissions. For this reason, a new set of scenarios were created by the scientific community and implemented in the more recent IPCC Fifth and Sixth Assessment reports, the so-called 'Representative Concentration Pathways' (RCPs).

- The A1 storyline describes of a world of rapid economic and population growth and rapid introduction of new and more efficient technology. Global population peaks in the middle of the century and declines thereafter. The major underlying themes are increased social and cultural interactions and convergence between regions with reduction in the differences in regional per capita income. The A1 family is split into three distinct scenarios: fossil fuel intensive—A1FI, non-fossil energy sources (more technological)—A1T, or a balance cross all energy sources—A1B.
- The A2 storyline describes a heterogeneous world where self-reliance and preservation of local identity is key. Population growth rates across all regions converge very slowly, leading to continuously increasing population. Technological advancement is slower and more fragmented than other storyline and economic development is primarily associated with regional and per capita economic growth.
- The B1 storyline describes a convergent world with a population that peaks in the middle of the century and declines thereafter (as in A1). There are rapid changes in economic structures towards a service and information economy, with associated reductions in material intensity and the introduction of clean resource efficient technologies. The emphasis is upon global solutions to economic social and environmental sustainability and equity, without additional climate initiatives.

- The B2 storyline describes a world where the emphasis is on local solutions to economic, social and environmental sustainability. Global population continuously increases but a slower rate than A2, with an intermediate level of economic development. Technological advancement is more diverse and the B1 and A1 storylines. This scenario is orientated towards environmental protection and social equity, but at a more regional local level.

There are four RCPs each covering the period 1850–2100 with extensions (extended concentration pathways (ECPs)) formulated for up to 2300. The RCPs are named according to the radiative forcing level at 2100, i.e., RCP 2.6, RCP 4.5, RCP 6 and RCP 8.5. The RCPs represent a simpler set of scenarios compared to SRES instead of four main storylines, each augmented with a range of possible scenarios based upon different socioeconomic futures. The RCPs are simply related to a level and acceleration of radiative forcing to avoid ambiguity. The science behind them is actually more complex than was used for the SRES storylines. The four RCPs are consistent with certain socioeconomic assumptions and are anticipated to provide flexible descriptions of different possible social, economic, demographic and technological futures. Each RCP has been shown to be achievable via several different regional and global socioeconomic routes. In this way, these new RCP scenarios are devolved from socioeconomic factors to help avoid confusion and dismissal by the general public. For similar reasons, four pathways were chosen rather than three to avoid the perception that the middle option is best and the safest bet. Table 1.2 provides a summary of the key features of the different RCP scenarios.

The details and relevance of the different social, economic, demographic and technological aspects of can be hard to comprehend. Essentially, all of these aspects can be reduced down to anthropogenic CO_2-eq emissions and a level of radiative forcing. Figure 1.14 shows changes to atmospheric CO_2 concentration form 1958 to the present day, figures 1.16 and 1.17 extend this trend in atmospheric CO_2 for the different SRES and RCP scenarios up to the end of the century.

We can see that despite the new science and inclusion of more recent socioeconomic data, such as the rise of China and India as major economic powers and

Table 1.2. Summary table of the representative concentration pathways

	Greenhouse gas emissions	Agricultural area	Air pollution
RCP 2.6	Very low	Medium for cropland and pasture	Medium-low
RCP 4.5	Medium-low, very low baseline	Very low for cropland and pasture	Medium
RCP 6	High mitigation, medium baseline	Medium for cropland, very low for pasture (total low)	Medium
RCP 8.5	High baseline	Medium for both cropland and pasture	Medium-high

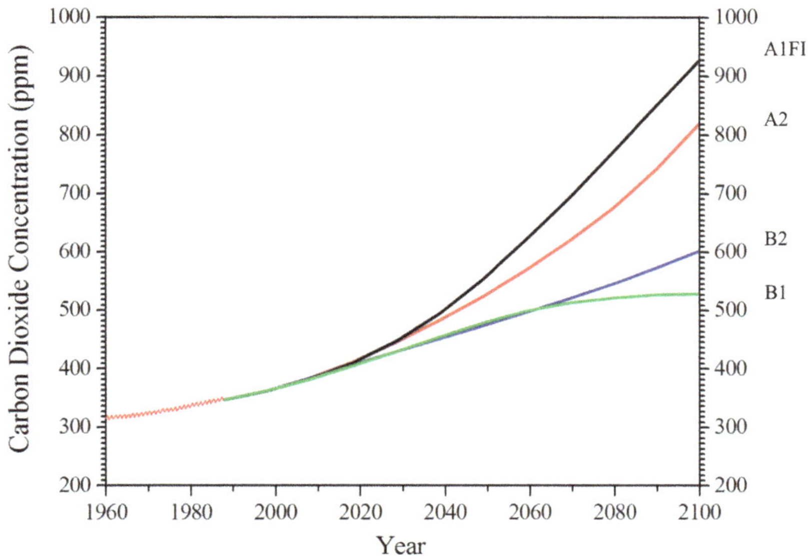

Figure 1.16. Global CO_2 concentration from 1960–2100 for four emissions scenarios overlaid with Mauna Loa historic CO_2 data. The range of uncertainty for each scenario is not shown.

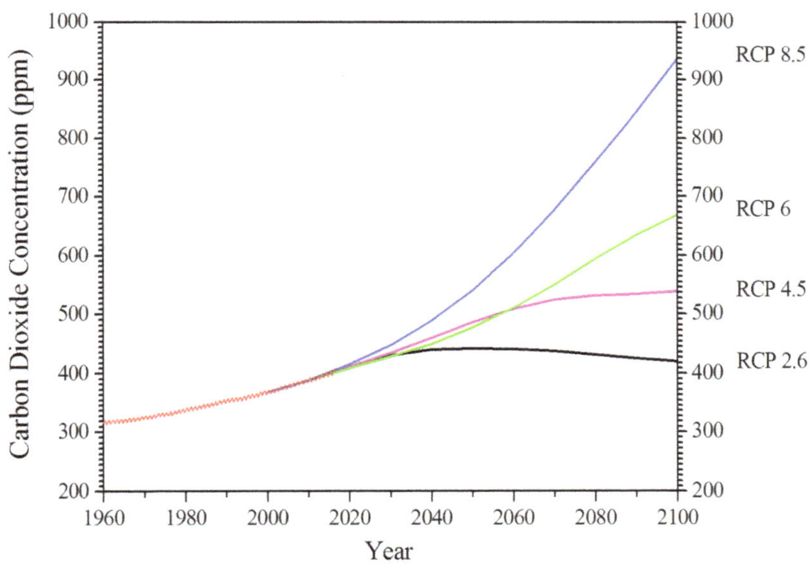

Figure 1.17. Global CO_2 concentration from 1960–2100 for the four RCPs, overlaid with Mauna Loa historic CO_2 data.

the associated emissions, there is little difference between the higher end scenarios (A1FI and RCP 8.5) by the end of the century. For brevity, the following chapters of this book will use climate change projections primarily based upon RCP 8.5, except in the case of specific products which are not available in the latest release of projection, in which case UKCP09 data based upon the SRES scenarios may be used.

The UK has among the most detailed climate change projections in the world. The current iteration of these projections, compiled by the Met Office Hadley Centre, were released in 2018 and are termed UKCP18 (UK Climate Projections 2018). These utilise the IPCC's Fifth Assessment report RCP scenarios and data but differ from many climate change projections in that they are probabilistic.

The UK Climate Projections 2018 are not solely based upon simulations of the UK Met Office Hadley Centre climate models, instead they are based upon many different climate models from around the world. This collection of climate models is run many times with different estimates for uncertain processes in the models (referred to as perturbations). Uncertainty in climate models arises from the natural variability in the climate system which is inherently random, our incomplete understanding of all of the Earth's climate processes as well as uncertainty in future greenhouse gas emissions. This allows the collection of climate models to create a spread of results based upon the limitations of our current understanding. By combining the results into distributions of possible values, or probability density functions (PDF), we can provide information about the relative likelihood of different future outcomes. This process is demonstrated graphically in figure 1.18 by comparing UKCP09 and UKCP18 to the previous iteration of UK climate projections UKCIP02.

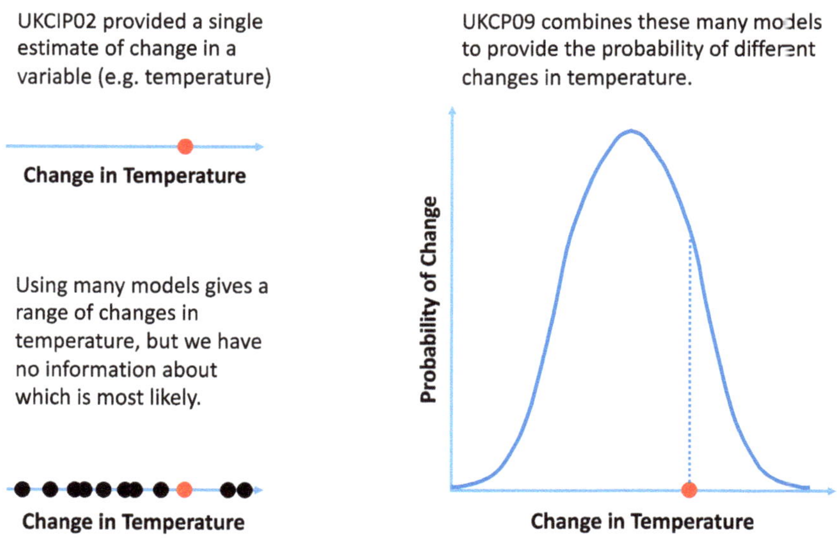

Figure 1.18. Illustration of how many climate model outputs are combined to form probabilistic estimate of climate change in UKCP09 (and UKCP18)[2].

---

[2] Based upon: Jenkins G J, Murphy J M, SextonD S, Lowe J A, Jones P and Kilsby C G 2009 *UK Climate Projections: Briefing Report* Met Office Hadley Centre, Exeter, UK. https://www.metoffice.gov.uk/research/approach/collaboration/ukcp

The current iteration of the UK climate projections UKCP18 consists of probabilistic data similar to UKCP09 (see figure 1.18) but at a finer spatial resolution, additionally there is greater emphasis on the representation and handling of extreme weather events such as storm surges within the projections.

1.6 Climate change impacts

Within this book, we talk about climate change and not global warming. Although the two terms are often used interchangeably in the media, this is not necessarily correct. The term global warming implies a global warming effect and could be interpreted to simply mean warmer temperatures, which those of us who live at higher latitudes might view as a good thing. However, the climate system is complex, the weather we experience is driven by atmospheric convection currents, the magnitude of which is a function of the rotation of the Earth and the difference in temperature between the surface and the upper atmosphere and between the equator and the poles. As some areas warm, other areas are expected to cool. The difference in temperature between the land and oceans will increase due to different thermal inertias altering the weather experienced in different global locations. Changes to these can disrupt and alter these convection currents, changing global weather patterns. Therefore, the term global warming is not strictly true, climate change is not just temperature but also sea level rise, changes to storms, monsoons, the persistence of droughts or the melting of permafrost to provide arable land. Figure 1.19 shows an illustration of the atmospheric convection currents, which drive global weather patterns, each hemisphere has three convection cells: the Polar cell, the Mid-latitude or Ferrel cell and the Hadley cell.

There is limited mixing between the atmospheric convection cells and air exchange might only occur during certain conditions. For example the Ozone hole in the Antarctic polar cell is a result of limited exchange between the southern Polar cells and the Ferrel cell during the cold Antarctic winter, the so-called Polar Vortex. Furthermore, we can see from figure 1.19 that at certain places air is rising or falling not travelling horizontally. It is for this reason that it was difficult for sailboats to cross the equator because air rises vertically here and there is little horizontal wind, giving rise to the 'doldrums'. Likewise, air is falling at the horse latitudes around 30° between the Ferrel cell and the Hadley cell. These atmospheric convection currents are also responsible for the trade winds at sea and the prevailing winds we experience on land.

There is a similar pattern at sea, the global ocean convection currents, which carry warm water near the surface from the tropics towards the poles where it cools and sinks, are responsible for the timing, magnitude and trajectory of monsoons in India, hurricanes in the Caribbean and storms in the UK. The temperature of the ocean surface determines all these things. The El Niño Southern Oscillation is the warming of the ocean surface in either the central and eastern tropical Pacific Ocean (La Niña refers to the cooling trend). This warming shifts the atmospheric circulation slightly, reducing rainfall over Indonesia and Australia, while increasing rainfall and monsoons over the tropical Pacific Ocean. Changes to ocean surface

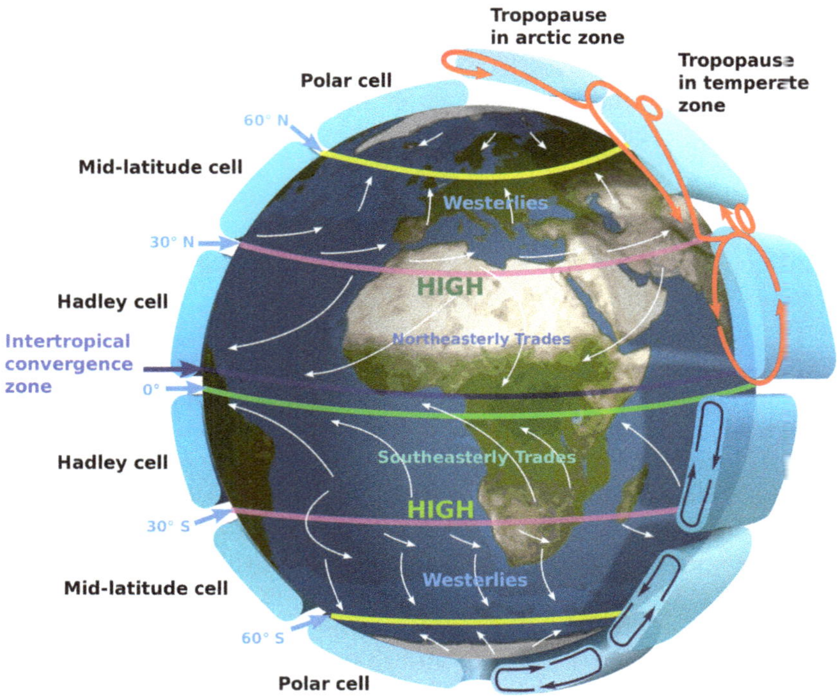

Figure 1.19. Illustration of the atmospheric convection currents that drive our climate system. This [Earth Global Circulation] image has been obtained by the author from the Wikimedia website where it was made available by [User:Kaidor] under a CC BY-SA 3.0 licence. It is included within this article on that basis. It is attributed to [User:Kaidor], adapted from a NASA image. https://commons.wikimedia.org/wiki/File:Earth_Global_Circulation_-_en.svg

temperature can have a dramatic effect on the weather experienced by different countries. A strong El Niño event recently produced a prolonged drought in Australia that lasted several years, threatening Australian wine production along the Darwin. The Atlantic Ocean also exhibits a similar effect, the Northern Atlantic Oscillation (NAO). The series of intense storms experienced during the summer of 2012 in the UK were as a result of a warmer than normal Atlantic surface temperature moving storm tracks southwards, which otherwise would have missed the UK and travelled on towards Iceland. It is important to note that climate change science has so far been unable to reach a consensus as to whether climate change will intensify or reduce the El Niño and NAO effects. However, it is clear to see then that changes to the status quo has the potential to have very profound influences on global weather.

The IPCC states that 'the resilience of many ecosystems is likely to be exceeded this century by an unprecedented combination of climate change, associated disturbances (e.g. flooding, drought, wildfire, insects, ocean acidification)'.[3] This is

[3] The quote can be found in the Summary for policy makers https://archive.ipcc.ch/publications_and_data/ar4/wg2/en/spmsspm-c-2-ecosystems.html

not just because of the magnitude of anticipated climate change but also the speed at which it will happen. We have seen previously that the Earth has experienced variation in temperature and climate before, but it is important to understand that these changes occurred slowly over millennia and ecosystems were able to adapt or migrate in response. The IPCC provides detailed estimates of the global impacts of a changing climate, some of which are summarised below.

Increased CO_2 in the atmosphere and warmer temperatures will lead to increased plant growth and hence increased CO_2 uptake. However, if temperatures rise too high, then plants begin to respire just like every other creature. Over the course of this century, net carbon uptake by terrestrial systems is expected to peak before the middle of the century and then weaken or even reverse, leading to amplification of climate change. It is estimated that 20%–30% of plant and animal species are likely to be at risk of extinction if the average global temperature increase is greater than 1.5 °C–2.5 °C. For temperature increases exceeding this range, there are expected to be major changes to ecosystems and the interactions between species and a migration of species. This will result in a reduction in biodiversity and will have impacts on water and food supplies.

Increased plant growth will lead to increased crop productivity at mid to high latitudes, with increased productivity depending upon temperature increase (1 °C–3 °C) and the crop in question. However, this will be balanced by reduction in crop productivity at lower latitudes. In seasonally dry and tropical regions, crop productivity is expected to decrease for even small increases in temperature (1 °C–2 °C). Globally, food production is projected to increase for a 1 °C–3 °C temperature increase but decrease thereafter.

Climate change will also have the potential to affect the health of millions of people. Reductions in crop productivity would increase malnutrition and increase the burden of diarrhoeal diseases. Changes to weather patterns resulting in more severe storms and extreme weather events increase risk of physical injury. Changes to weather and climate will alter the global distribution of infectious diseases. In urban areas increased concentrations of ground level ozone increases risk of cardio-respiratory diseases. There will be benefits as well, in temperate zones such as the UK, cold related deaths in winter currently far outweigh heat related illness in summer. Climate change will reduce the number of cold related deaths in these areas, but higher summertime temperatures and increased instances and intensities of heat waves will result in more deaths in summer. In the short term, temperate areas should see reduction in temperature related deaths. Globally, however, rising temperatures will result in increased mortality, especially in developing countries.

Changes to the frequency and intensity of precipitation will lead to increased stress on water resources, which will be exacerbated by population growth and land-use change such as urbanisation. Changes to precipitation and temperature will increase surface water run-off. In particular, run-off is projected to increase by up to 40% at higher latitudes and in some wet-tropical areas (e.g. Southeast Asia) but at the same time decrease by up to 30% over dry areas at the mid-latitudes and in dry-tropical areas, due to decreased rainfall but also increased uptake and evapotranspiration by plants. Areas that are currently affected by drought are expected to grow in size, having an adverse impact upon agriculture, water supply and human health. Climate research suggests that there will be a significant increase in future heavy rainfall events for many regions, even regions where the mean total annual rainfall is projected to decrease. This poses challenges to society in the form of floods, damage to infrastructure and water quality.

Sea level rise due to thermal expansion of the oceans as they warm will threaten many low-lying coastal areas. The greatest numbers will be in the densely populated mega-deltas of Asia and Africa, while small islands are particularly vulnerable and in danger of disappearing completely (e.g. Micronesia). In addition, thermal expansion will increase ocean height due to stronger storms and greater storm surges. In the presence of a low pressure system, the atmosphere exerts less force on the ocean surface and the level rises. Storm surges in the UK can currently exceed 1 m in height, stronger storms will be typified by even lower pressure systems and hence greater storm surges. This will pose particular problems for low-lying areas, especially those with dense urban populations. The most vulnerable societies, industries and settlements are those in coastal and river flood plains, especially those where the economy is dependent on climate sensitive resources and sensitive to extreme weather events (e.g. harbours). Rapid urbanisation in these areas will exacerbate the risk from climate change.

The further chapters of this book deal with specific climate change risks and their impact on urban areas. As might be expected, there will be some overlap and common solutions to some of the climate change related issues. The final chapter aims to summarise these and tease out some common solutions to increase the resilience of urban areas to climate change.

IOP Publishing

Climate Change Resilience in the Urban Environment (Second Edition)

Tristan Kershaw

Chapter 2

Decarbonisation and mitigation targets for the built environment

Buildings are responsible for over 30% of global energy consumption, which when you consider how this energy is created, equates to more than 25% of global energy-related CO_2 emissions. The global buildings sector is therefore an area of intense interest when it comes to the mitigation of climate change. However, many countries lack building energy efficiency regulations, and the next 20 years will see a rapid increase in the global building stock, predominantly in these countries.

Luckily, the global energy system is evolving at a rapid rate. In 2016, renewable overtook coal as the largest source of energy generation in terms of capacity. The rapid and aggressive deployment of clean energy technologies could put the world on the path to a carbon-neutral energy system by 2060, this is crucial in order to limit climate change. Limiting global average temperature increase to 1.75 °C above pre-industrial levels by 2100, the mid-point of the 2015 Paris Agreement range, is technically feasible. However, there is a huge gap between this pathway and our current efforts, indicating that an acceleration of action at a global level is required. This chapter will consider our current efforts and the barriers to meeting these targets, as well as possible solutions.

2.1 The problem with renewables

As a child I grew up looking over the Solent at Fawley power station near Southampton. This oil-fired power station was not in constant use and instead, being relatively quick to turn on and off, was used to meet periods of peak demand. Things such as the football World Cup final would see this station fired up to meet the surge in demand. This now demolished power station was a reminder that our energy consumption is not constant, and that our generation needs to rise or fall to meet demand.

doi:10.1088/978-0-7503-5262-8ch2

The amount of primary energy used in the UK has been falling since the year 2000 from 234.8 Mtoe (Million tonnes of oil equivalent) to 170.1 Mtoe in 2021, or equivalently 1978 TWh. This equates to a final energy consumption of 128.2 Mtoe or 1491 TWh[1]. These numbers include directly consumed fuels such natural gas, petrol and biomass (see figure 2.1). According to the Office of National Statistics, in 2021 the UK used 318 TWh of electricity, or if we assume a constant use, a demand of ~36 GW. The UK's current installed generation capacity though is nearly triple that figure at 105 GW in order to meet the variations in demand and provide sufficient resilience. Herein lies the largest hurdle we face when decarbonising the energy supply to curb carbon emissions and the resultant climatic change—we need to meet not only our average energy consumption, but also our peak consumption.

In order for a renewable energy source to be useful it needs to be converted to a suitable form (e.g. electricity or heat) and it needs to be available when required. It is this second point that becomes more poignant as countries strive to reduce their reliance on fossil fuels and move more towards renewable energy output. On the face of it, renewable energy systems such as photovoltaics, wind turbines, hydroelectric and tidal energy systems seem perfect, once installed they produce 'free' energy for years to come. The glitch though is that we cannot just turn the Sun or wind on or off. We have no control over the weather and cannot accurately predict future

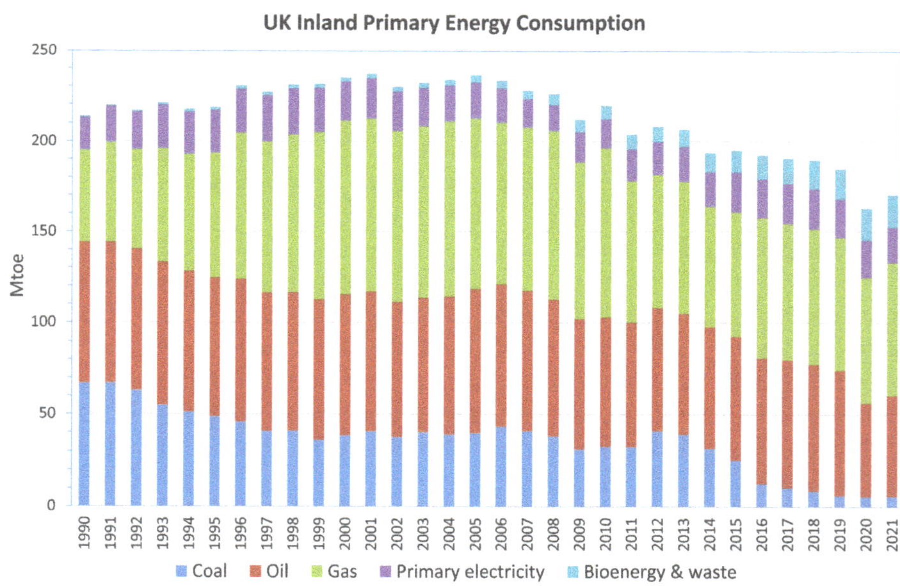

Figure 2.1. Plot of the UK's primary energy consumption over time broken down by fuel type. Data obtained from 'UK Energy in Brief 2022' under Open Government Licence v3.0.

[1] Primary energy is the raw energy harvested from natural resources before any transformation or losses associated with transformation (e.g. electricity generation) or transmission losses. Final energy is the delivered or usable energy (e.g. the energy you pay for).

generation. Tidal energy at least is predictable—we have relatively accurate knowledge of the size and timing of tides and if we ignore ecological considerations for a moment, it is possible to store this energy for when we need it. Similarly, with hydroelectric energy we have some control over the energy generated and the timing, but we are ultimately at the mercy of when it rains, and the amount of water stored.

From figure 2.2 we can see that the UK's electricity generation capacity has increased over time, mostly as a result of increased renewable generation capacity. We can expect that over the next few decades the UK and other countries will move towards an all-electric energy supply system, phasing out the direct consumption of coal, oil and gas by users. This allows for greater integration of renewable energy but will increase the demand upon the energy supply system, as mentioned earlier the UK used 318 TWh of electricity in 2021, which was supplied by a variety of fuel types (see figure 2.3). We can see that over time the fuels that make up the UK's grid electricity have changed, and hence so has the carbon content of a unit of electricity. In order to meet climate change mitigation targets we will need to continue reducing the carbon content of our grid electricity, which implies more renewable energy. However, if we move to an all-electric supply, then the capacity will need to increase to supply all the 1491 TWh (128 Mtoe) of final energy consumed in 2021 (figure 2.1) and not just the current 318 TWh. This would mean increasing our electricity supply from ~36 GW currently to ~170 GW and the total generation capacity to over half

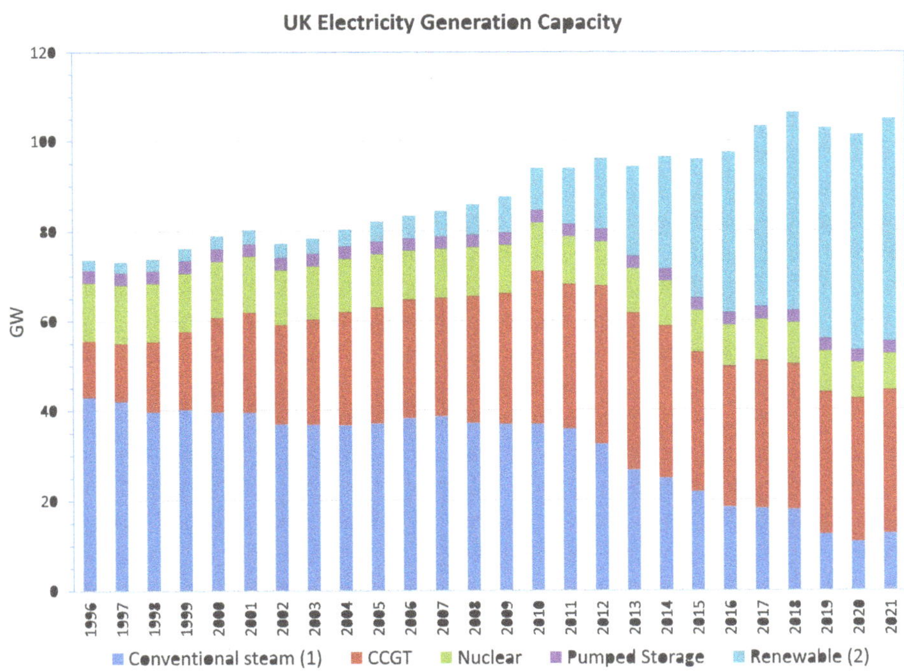

Figure 2.2. Plot of the UK's electricity generation capacity over time, broken down by type (CCGT = combined-cycle gas turbines). (1) Includes coal, non-CCGT gas, oil and mixed/dual fired, but does not include thermal renewables. (2) Renewable capacity is on an installed capacity basis. Data obtained from 'UK Energy in Brief 2022' under Open Government Licence v3.0.

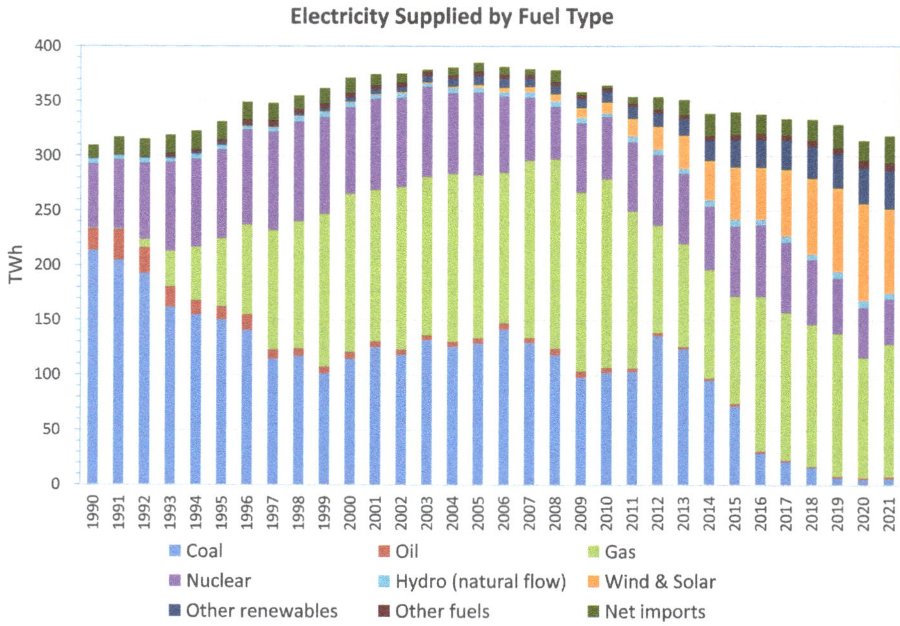

Figure 2.3. Breakdown of UK electricity supply by fuel type over time. Data obtained from 'UK Energy in Brief 2022' under Open Government Licence v3.0.

a TW. We can expect that this electricity generation, which currently consists of a mixture of fuel types, will be replaced with a mixture of wind, solar and hydroelectric renewables, augmented with nuclear power; this however, is a mammoth task. Unlike fossil fuels and nuclear power, renewables are by their very nature variable. Some of this variability is obvious, tides come in and out, and the Sun sets at night. Some variation, however, is less obvious, there is seasonal variability of Sun, wind and rain, but also daily, hourly and even minutely variation in these resources that become almost impossible to accurately predict. One of the greatest challenges then is the ability to store renewable energy for those periods of variation when the wind is not blowing, and the Sun is not shining. Factor in the periods of peak demand that currently require a factor of three increase in generation capacity above average use and there is a clear issue that needs to be addressed.

The choice of storage medium requires some thought. Pumped storage has long been a mechanism for storing surplus electricity for when it is required. However, if we look at figure 2.2, then we can see that the capacity of pumped storage has not increase over time to match the increase in renewable energy capacity. Pumped storage which converts surplus electricity to potential energy requires a reservoir at height in which to store this potential energy (PE), which would be given by:

$$\mathrm{PE} = mgh$$

where m is the mass of the water, g is acceleration due to gravity (9.8 m s^{-2}) and h is the height through which the water is travelling. When the water flows, this energy is converted to kinetic energy, which is then harvested by a turbine and converted back into electricity. This system is not without losses due to friction, and the pump and turbine efficiencies. However, the greatest issue is the availability of suitable sites. Projects such as the Three Gorges Dam or the Hoover Dam require significant land and infrastructure, and the potential removal of settlements and the destruction of the existing habitat. The number and scale of such projects to store sufficient energy for our needs makes pumped storage on its own unviable.

So how much storage will we need? We will need sufficient storage capacity to cover periods where renewable generation in minimal, those doldrum periods where solar radiation is at a minimum and the wind is not blowing. We will also need sufficiently increased renewable generation capacity prior to this doldrum event to provide energy to be stored. On a cold winters day (for example) the UK uses ~1 TWh of electricity, or equivalently about 40 GW. Currently half of this electricity is supplied by renewable energy (figures 2.2 and 2.3). We need then to store at least 0.5 TWh currently because the remainder can be covered through other means. This poses some philosophical questions: Are structured blackouts or brownouts acceptable during such periods? And, will the public accept this? How much storage is required therefore depends upon the number of possible days that renewable generation could be diminished and what level of risk is deemed acceptable when contingency planning. If we (for example) consider a 10 day doldrum period when renewable generation is reduced, something that we in the UK are quite used to, we would need at least 5 TWh of renewable energy storage capacity. If we assume that this will be met through pumped storage, then we would need 23 000 $m^3 s^{-1}$ of water flowing constantly for 10 days to generate the ~20 GW needed (assuming a 100 m head and 90% efficiency). For comparison, the typical flow rate over Niagara Falls is only 2400 $m^3 s^{-1}$. It seems clear that other methods of storage are needed in order to meet our storage requirements, one such medium is Lithium-Ion battery technology. Lithium-Ion batteries are now relatively commonplace in mobile phones and electric vehicles; however, their capacity is typically small, a common electric vehicle battery currently only holds around 80 kWh. The benefit of battery technology is that it stores electrical energy, and therefore the energy does not need to change form. Currently, a 1 GWh Lithium-Ion battery storage facility is being constructed in London. This project is estimated to cost around £300 m to construct, so scaling such a facility up to meet our needs would imply a £1.5 trillion price tag. This, however, is not the entire story. As we move towards a decarbonised all-electric energy supply based entirely upon renewables, we need to consider how we would store sufficient energy not just for 50% of our current electricity use but for 100% of our final energy demand, which for a 10-day period would be ~40 TWh, which following the same calculation could cost ~£12 trillion. While we can expect prices to drop, or that another cheaper battery technology may be developed, it seems more likely that energy will need to be stored in a different way than either pumped storage or Lithium-Ion batteries. One such possibility is the electrolysis of water or ammonia to create hydrogen, which can either be converted back to electricity via combustion or directly through the use of fuel cells. The sufficient generation and storage of hydrogen poses another set of problems and this technology has not yet reached sufficient maturity, although much research is currently ongoing.

This section has focused principally on the UK for simplicity, but these issues are prevalent worldwide. Net-zero carbon targets will require almost all our energy to be provided by renewable electricity in the near future. It is therefore essential that we find ways of generating and storing sufficient surplus of renewable energy for doldrum periods. Currently, this rather obvious issue is being neglected by governments around the world. The technology and space constraints make the generation and storage of sufficient energy unrealistic based upon our current usage. The only remaining choice is therefore to improve our efficiency levels and reduce the amount of energy that we consume, and therefore the amount that needs to be generated and stored.

2.2 Increasing efficiency in the buildings sector

Despite the consensus on climatic change and the global drive to reduce carbon emission, there is a dearth of sector specific targets for countries to aim towards. The global buildings sector is responsible for over 30% of the worlds final energy consumption and 28% of the worlds carbon emissions. The vast majority (80%–90%) of these emissions are from buildings in use, rather than construction. It is therefore this operational energy use over the lifetime of a building, which comprises of heating, cooling, ventilation, lighting and plug loads, that needs to be addressed to have the greatest effect when trying to reduce carbon emission from the buildings sector. The International Energy Agency (IEA) is one of the few organisations that offer specific targets for countries to aim towards to mitigate carbon emissions from the buildings sector.

The IEA's 2 °C Scenario (2DS) sets out an energy use and carbon emissions pathway consistent with at least a 50% chance to limit global average temperature increase to 2 °C by 2100. This is consistent with reducing carbon emission from the energy system by 70% by 2060, and to reach carbon neutrality by 2100. The 2DS follows a rapid decarbonisation pathway and is based upon an ambitious uptake of renewable energy and the phasing out of fossil fuels. However, as discussed in the previous section, this scenario does recognise the limitations of renewable energy and the current capability to store renewable energy. As such, the energy mix of the 2DS pathway includes a diverse mix of energy generation technologies, including nuclear and fossil fuels with carbon capture. The upshot of this is that in order to meet the carbon emission targets, severe efficiency gains need to be made across various energy-use sectors, including the buildings sector.

This, however, is not as simple as it seems. While the global focus has been to reduce operational carbon emissions from buildings through the decarbonisation of the energy supply system and the implementation and tightening of the regulations and standards, we are yet to see tangible reductions in the buildings sectors total energy consumption. It has been estimated that building energy efficiency measures have saved more than 450 Exajoules (EJ) in cumulative energy savings over the last 25 years. This amounts to a significant amount of un-emitted carbon; however, the potential for efficiency savings remains largely untapped. Data presented by the IEA shows that global consumption continued to grow by nearly 2% per annum over the 40-year period up to 2010. This was responsible for nearly a doubling of the associate carbon emissions over the same period. The reasons commonly attributed to this continuing increase are population and socioeconomic growth. An increasing population requires a greater built floor area to house it, while an increasing gross domestic product (GDP) per capita leads to greater expectations of comfort (heating and cooling energy usage) and greater expenditure on products and services (e.g. more energy consuming appliances). There is also the less expected result that as people's income increases, they are able to afford larger properties, and so the floor area per capita also increases. By 2060, income per capita in developing countries is expected to more than quadruple and the associated global floor area will more than double. About half of the buildings that will be built between now and 2060, over 100 billion m^2, will be built in countries that currently have no mandatory building energy codes in place. The next 20 years will see over half of new additions to 2060's building stock being built, and by 2035 nearly two thirds of the global buildings stock to 2060 will already be standing. This means that immediate steps need to be taken to avoid locked-in building energy inefficiency. This is not just an issue in developing countries, even countries like the UK struggle to deploy energy efficiency measures. Figure 2.4 shows the number of homes with wall or roof insulation. Given that there are ~27 million homes in the UK, there is still a significant number of homes without energy efficiency measures, despite increasing energy prices, Government grants and advertising campaigns to promote awareness.

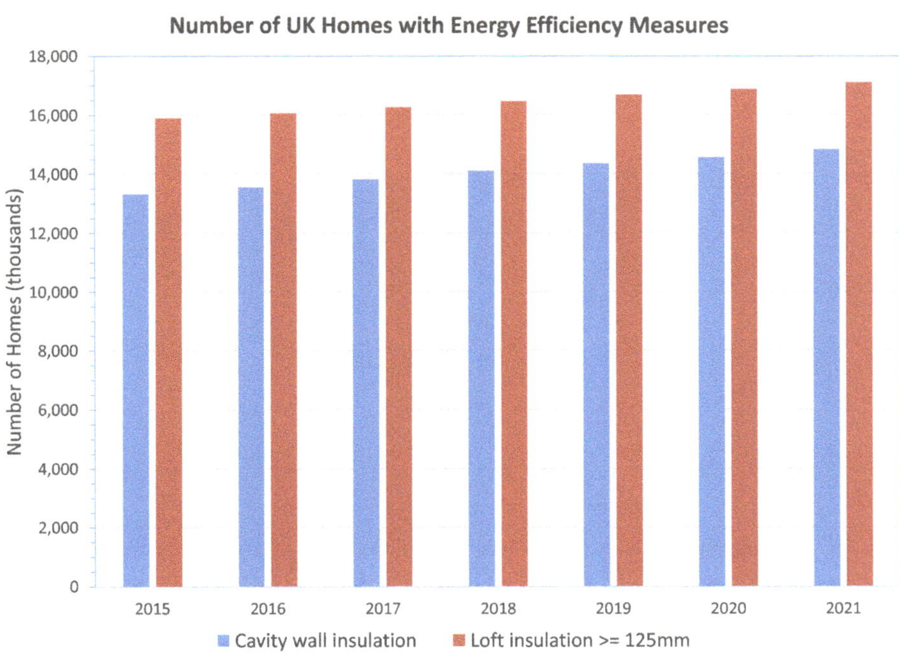

Figure 2.4. Number of UK homes with cavity wall insulation or loft insulation. Data obtained from 'UK Energy in Brief 2022' under Open Government Licence v3.0.

The global buildings sector consumed ~125 EJ in 2014, or about 30% of the global final energy demand and 55% of the final electricity demand. When the carbon emission of the various energy types is taken into consideration, it can be seen that buildings are responsible for over a quarter of global energy-related CO_2 emissions. In the IEA's Energy Technology Perspectives (2017) report, the IEA states that under the 2DS pathway the buildings sector needs to limit its final energy consumption to 130 EJ (36.1 PWh)[2] by 2045 and remain relatively constant thereafter to 2060. Another target is also mentioned, which is limiting per capita final energy usage to greater than 3.5 MWh/person/year by 2060. The following sections will examine estimates of population and socioeconomic growth to explore the likelihood of the buildings sector meeting the 2DS target of the IEA.

2.3 Projecting global buildings sector energy consumption

This section will perform some simple analysis of how likely we are to meet our climate change targets for the buildings sector using a dynamic stock-driven model (i.e. a model that is dependent upon changes in population and GDP over time). The World Bank provides various estimates of future population growth based different factors[5]. For simplicity, we are only going to consider the high, median, and low estimates of population growth. Figure 2.5 shows historic global population to 2020

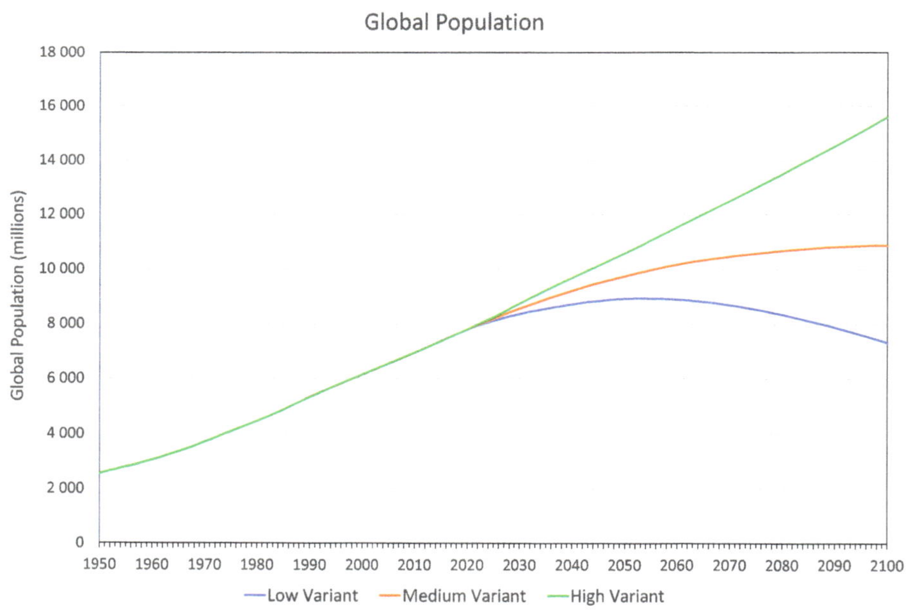

Figure 2.5. Historic and estimates of future global population growth. Data taken from the World Bank[3].

[2] Note on units used in global energy consumption: 1 Exajoule = 10^{18} J, 1 PWh = 1000 TWh or 10^{12} kWh.

[3] United Nations, Department of Economic and Social Affairs, Population Division (2019). World Population Prospects 2019, Online Edition. Rev. 1. CC BY 3.0 IGO https://population.un.org/wpp2019/

and the three variants thereafter. As we can see, there is considerable disparity between these population growth trends, which will have a profound effect on the global final energy consumption.

While figure 2.5 shows the total global population, the World Bank datasets provide population data and estimates for 235 countries and territories, which is used in this analysis. Estimates of GDP per capita and growth over time are, however, harder to find. The work of Harvey (2014) forms the basis of this study, where the world was split into 10 representative regions, each with estimates of economic growth that can be used to estimate future global floor areas. These 10 regions are based upon those used by the IPCC in its Special Report on Emissions Scenarios (SRES), the 235 countries are each attributed to one of the regions. The SRES regions are Pacific Asia OECD (PAO), North America (NAM), Western Europe (WEU), Eastern Europe (EEU), Former Soviet Union (FSU), Latin America (LAM), Sub-Saharan Africa (SSA), Middle East and North Africa (MENA), Centrally-planned Asia (CPA) and South and Pacific Asia (SPA).

Harvey's model starts in 2010 and runs to 2100. Given that we are using the parameters presented in the paper, we have done the same here, albeit updating the 2010 recorded population data with the World Bank's latest estimates. The model considers population, GDP per capita and residential and commercial floor area, together with how all of these are expected to change over time for the 10 different regions. Two scenarios of economic growth are presented, low and high, represented by a different asymptotic GDP G_∞ (i.e. a maximum value that GDP will approach but never exceed).

Table 2.1 shows the starting conditions for the economic part of our stock model. As we can see, each region has both a different starting GDP, different growth parameters and a different asymptotic GDP, implying that even by 2100 there will

Table 2.1. Initial population (2010), starting GDP per capita (G_0) and growth parameters (G_∞ and α) for the 10 world regions. Data sourced from World Bank and Harvey (2014).

		GDP (US $/person)			
			Asymptotic (G_∞)		
	Population (millions) 2010	2010 (G_0)	Low	High	Growth parameter α
PAO	205	33 990	34 000	50 000	0.02
NAM	347	45 516	40 000	55 000	0.01
WEU	484	30 923	30 000	50 000	0.02
EEU	117	16 489	30 000	45 000	0.04
FSU	288	11 412	27 000	45 000	0.04
LAM	588	10 966	25 000	40 000	0.04
SSA	836	2232	22 000	30 000	0.04
MENA	420	8930	22 000	35 000	0.04
CPA	1512	7270	27 000	45 000	0.06
SPA	2161	4127	27 000	40 000	0.06

still be economic disparity between countries and regions. We can estimate the GDP per capita per region at year t using the following formula and the values in table 2.1.

$$G(t) = \frac{G_\infty}{1 + \left(\frac{G_\infty - G_0}{G_0}\right)e^{-\alpha(t-t_0)}}$$

where t_0 is the starting year, in this case 2010. As time progresses, $G(t)$ will asymptotically approach G_∞. The amount of residential and commercial built floor area is assumed to increase not only with increasing population but also increasing GDP per capita. The future floor area per person is given the following formula and the values in table 2.2.

$$A(t) = \frac{A_\infty}{1 + \left(\frac{A_\infty - A_0}{A_0}\right)e^{-\beta(G(t)-G_0)}}$$

Having now estimated how the amount of floor for both residential and commercial buildings will change for each region up to the end of the century, we now need to convert this to an estimate of energy usage. The energy use intensity (EUI) measured in kWh m^{-2} is a common benchmark for assessing the efficiency of a building. This, however, will likely change over time as the different regions undergo socioeconomic change. Once we have an estimate of the average EUI of the built stock for each year, it is simply a matter of multiplying that by the floor area per person and the population in order to obtain a total energy consumption (EC). This is achieved using the following equation and the values presented in table 2.3.

Table 2.2. Initial floor areas (2010) and growth parameters (A_∞ and β) for the 10 world regions. Calculated using data from World Bank and Harvey (2014).

	Residential buildings			Commercial buildings			
		Per capita floor area (m^2)			Per capita floor area (m^2)		
	Floor area (billion m^2)	2010 (A_0)	Asymptotic (A_∞)	Floor area (billion m^2)	2010 (A_0)	Asymptotic (A_∞)	Growth parameter β
PAO	8.1	40	55	3	14.7	25	0.0002
NAM	22.6	65	60	8.3	23.9	25	0.0002
WEU	17.5	36	45	6.9	14.3	20	0.0002
EEU	2.9	25	45	0.9	7.7	18	0.0002
FSU	7.0	24	45	1.9	6.6	18	0.0002
LAM	9.4	16	40	2.7	4,6	18	0.0002
SSA	8.0	10	40	1.1	1.3	15	0.0002
MENA	6.9	16	40	1.7	4.1	15	0.0002
CPA	41.4	27	40	8.0	5.3	15	0.0002
SPA	28.1	13	40	4.8	2.2	15	0.0002

Table 2.3. Initial EUI (2010) and growth parameters (EUI_∞ and β) for the 10 world regions. Data sourced from [1].

	Residential buildings ($kWh\,m^{-2}\,yr^{-1}$)		Commercial buildings ($kWh\,m^{-2}\,yr^{-1}$)		
	2010 (EUI_0)	Asymptotic (EUI_∞)	2010 (EUI_0)	Asymptotic (EUI_∞)	Growth parameter β
PAO	108	169	357	360	0.0002
NAM	148	151	347	353	0.0002
WEU	185	187	251	278	0.0002
EEU	202	203	306	325	0.0002
FSU	256	260	382	411	0.0002
LAM	52	77	125	185	0.0002
SSA	189	215	86	213	0.0002
MENA	122	155	204	261	0.0002
CPA	75	160	90	199	0.0002
SPA	72	113	112	254	0.0002

$$EC(t) = \frac{EUI_\infty}{1 + \left(\frac{EUI_\infty - EUI_0}{EUI_0}\right)e^{-\beta(A(t)-A_0)}} \times A(t) \times POP(t)$$

where POP(t) is the region's population at year t.

As we can see in table 2.3, there is currently a large disparity in the amount of energy used in buildings across the different regions. This disparity will become smaller over time but we are unlikely to achieve equality by 2100 according to the current estimates of Harvey and the World Bank. We have used a constant growth parameter of 0.0002 for the EUI, the same as for the floor area, because we have assumed that the EUI will grow with floor area as new buildings are erected. A more aggressive scenario would be to assume that EUI grows in-line with GDP/capita and uses the GDP growth parameter α. This analysis was ultimately unnecessary, however, because figure 2.6 shows that even the more conservative scenario fails to meet the IEA's 2DS target of 36.1 PWh (130 EJ) in 2045.

Figure 2.6 shows the business-as-usual case using the population estimates from the World Bank and the socioeconomic projections of GDP, floor area and EUI growth from Harvey (2014). Six scenarios are presented, representing the three population growth estimates (see figure 2.5) and the high and low GDP/capita growth estimates (see table 2.1). What is concerning is not only that this business-as-usual case misses the 2045 target but it also fails to meet it later this century, even under the low population and low growth scenarios. Remember that the final energy consumption was meant to peak in 2045 and remain relatively constant to 2060 (the IEA 2DS scenario ends at 2060). This highlights just how far the buildings sector has to go in order to meet its contribution to global climate change targets.

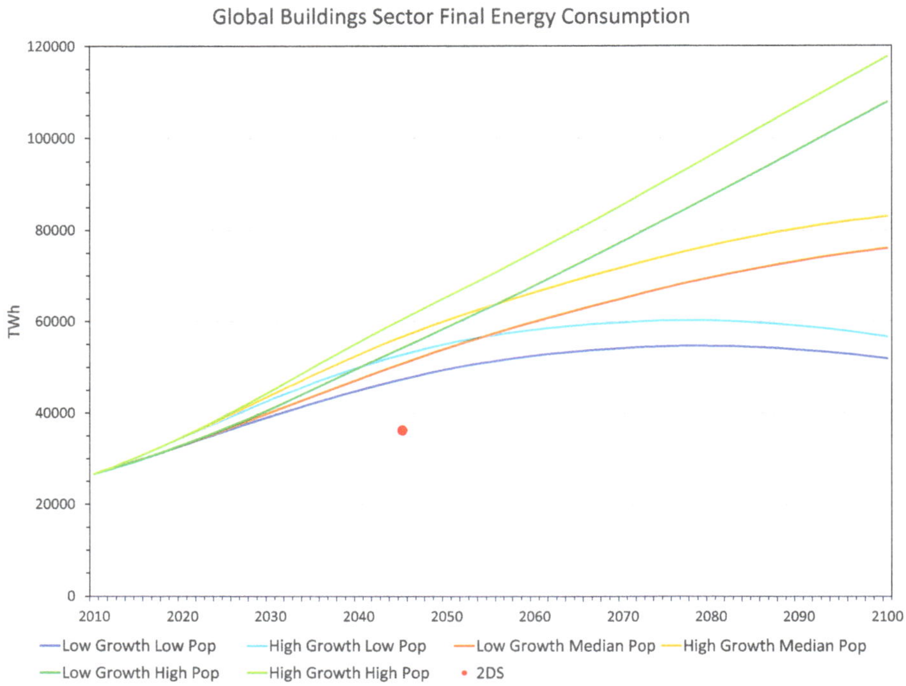

Figure 2.6. Business as usual projections of global building energy use based upon different estimates of economic and population growth.

2.4 Can current energy efficiency codes save the planet?

Now that we have a model capable of projecting building sector energy consumption over time, we are now able to consider some what if scenarios. The EUI values presented in table 2.3, while realistic, are nowhere near best-practice levels. There are numerous building energy efficiency standards available, the most common standard is the Leadership in Energy and Environmental Design (LEED) standard offered by the U.S. Green Building Council. The energy part of LEED is based on a reduction over the American Society of Heating, Refrigerating and Air-Conditioning Engineers (ASHRAE) building codes. The popularity of the ASHRAE codes globally has meant that LEED is the most widespread energy efficiency standard. LEED, like the Building Research Establishment Environmental Assessment Methodology (BREEAM) in the UK, is a sustainability standard. Points are awarded for a variety of different environmental and sustainability criteria, ranging from biodiversity to reducing embodied carbon and air-quality to promoting recycling on-site, building energy use is only one part of this assessment.

Another popular building energy efficiency standard is Passivhaus. Originating in Germany the Passivhaus standard[4] is probably the strictest energy efficiency standard available currently. Unlike LEED which is linked to the ASHRAE EUI for

[4] For more details, please see https://www.passivhaustrust.org.uk.

different climate zones, Passivhaus sets a single limit of 60 $\mathrm{kWh\,m^{-2}}$ for all building types regardless of use. This 60 $\mathrm{kWh\,m^{-2}}$ is primary energy and different factors are applied to convert from final energy to primary energy depending on the location and fuel type. For simplicity, here we will use a conversion factor of 1.00 (i.e. primary energy = final energy), which would be indicative of generating all energy on-site. Even with this rather generous assumption, the EUI of a Passivhaus is still lower than any of the regional average EUIs presented in table 2.3 if we take our business-as-usual case shown in figure 2.6 and from the year 2020 assume that all new buildings are constructed as Passivhaus certified with an EUI of 60 $\mathrm{kWh\,m^{-2}}$. Additionally, we will assume that 2% of the existing building stock in 2019 is either demolished and rebuilt or retrofit to meet the Passivhaus standard. This results in almost all the currently existing building stock being replaced by 2100. The results of this exercise can be seen in figure 2.7.

Figure 2.7 shows that even with all new buildings from 2020 constructed to our most stringent building energy standard, only half of our scenarios meet the 2DS target of <36.1 PWh in 2045. The estimates of population growth are the dominant factor here. The low population growth variant shows a steady decline in global buildings sector energy consumption after 2045. Whereas the high population growth variants continue to grow and are still accelerating at 2100. The median growth estimates are perhaps the most interesting. While the World Bank prescribes no likelihood to these different estimates, this middle of the road estimate is likely to be close to reality (if we assume a normal distribution). For the median variant we

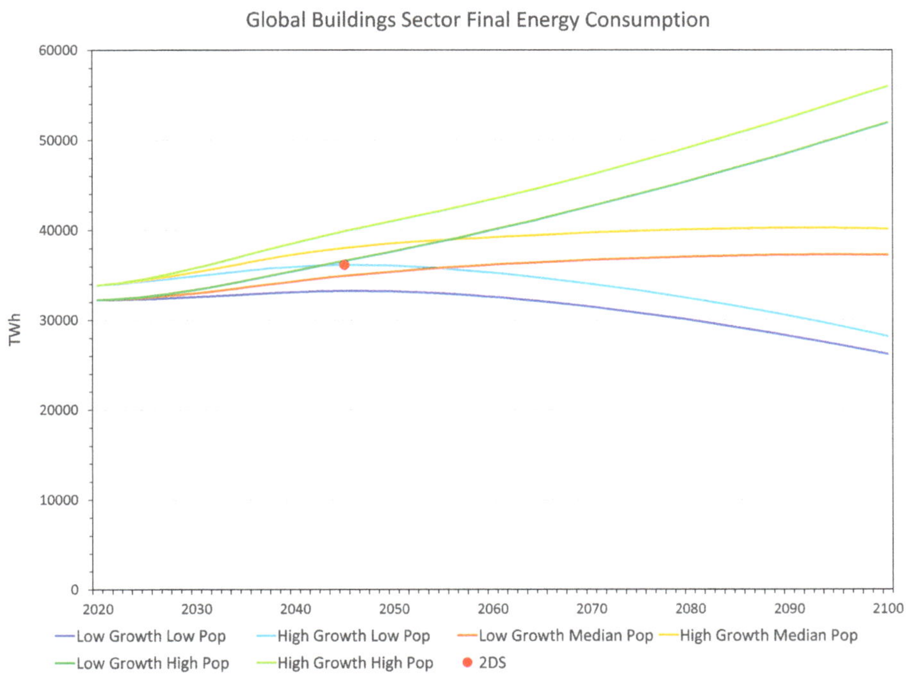

Figure 2.7. Plot of global building energy use based upon all new buildings' Passivhaus levels of efficiency.

see that the low economic growth version passes the 2DS test while the high economic growth version just fails to meet the target. However, both of these scenarios remain relatively constant after 2045 up to 2060 and beyond, as stipulated by the IEA.

If we consider the alternate energy use per capita target stipulated by the IEA of 3.5 MWh/person/year by 2060, we have a very different picture. Figure 2.8 shows the results of normalising the data shown in figure 2.7 with the relevant population growth variant. As we can see, only the high population variant with low economic growth passes this criterion. The median population variant again with low economic growth just misses the target but meets it by ~2070.

This analysis does not paint a very encouraging picture. Even with our current best-practice levels of efficiency, we are unable to consistently meet the 2DS targets. We also need to remember that this analysis, while simplified, is highly optimistic. The IEAs 2DS target is based upon the IEAs assumptions on the widespread uptake of renewable energy. The IEA assumes a global average decarbonisation rate of 2.8% per year. This is more than four times greater than the highest rate ever recorded. The IEA assumes that the energy sector will be made up of renewables (both on-site and off-site), nuclear and some fossil fuels with carbon capture (i.e. not all our energy demands can be met through entirely renewable means). We have used the Passivhaus standard with an EUI of 60 $kWh\,m^{-2}$, assuming all on-site generation with no losses so that the primary energy is equivalent to the final energy.

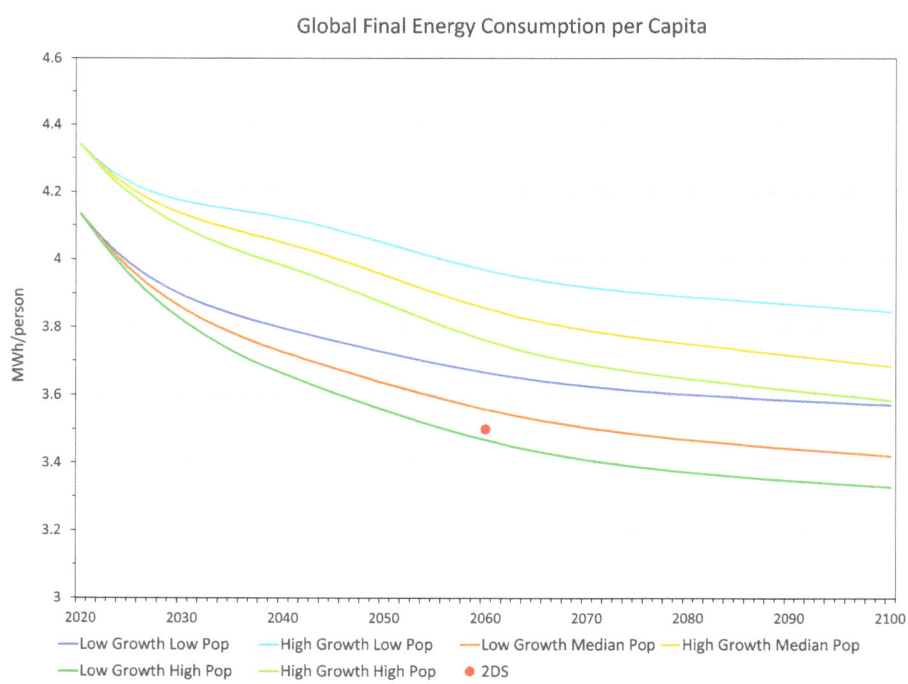

Figure 2.8. Plot of global building energy use normalised by global population, based upon all new buildings' Passivhaus levels of efficiency.

In reality there will be losses associated with generation, transmission and storage/ conversion of energy, in-line with the IEA's assumptions, this is not reflected in the graphs. The biggest assumption is that all new buildings will be constructed to Passivhaus levels of efficiency. Passivhaus, while gaining notoriety, is nowhere near as popular as LEED in terms of global distribution and sheer numbers. While it is not simple to directly compare to Passivhaus to LEED because the LEED's target EUI varies with the background climate, a typical EUI for LEED Platinum (the highest award available) would be 90 $kWh\,m^{-2}$ (i.e. 50% worse than Passivhaus).

One factor that is hard to consider in the analysis is what is called the rebound effect, whereby energy (or monetary) savings from increased efficiency are not realised in practice. For instance, increasing insulation levels in a home would typically save energy; however, this is combatted by turning up the thermostat and increasing the heating set-point temperature. The efficiency of our devices has increased tremendously over the last decade and yet we have combatted this by buying more devices. This rebound effect can take many forms but ultimately it acts as a barrier to realising efficiency gains and limiting global energy use. Various studies have measured the rebound effect and found that it can vary considerably, from a few percent up to 92% of predicted efficiency gains not being realised.

2.5 Conclusions

Given the eye-opening analysis shown in this chapter, we are faced with a sobering reality in which in order to avoid dangerous levels of climate change our buildings sector needs to dramatically increase its levels of efficiency. Even a stepwise change to current best practice is insufficient to meet the targets set out in the IEA's 2DS pathway, namely limiting global energy consumption to 130 EJ (36.1 PWh) by 2045 and remaining constant to 2060 (and reducing thereafter), or reducing energy use per capita to 3.5 MWh/person/year by 2060. It is clear that we need to increase efficiency levels beyond current best practice and/or we need to exceed the IEA's estimates for available renewable energy generation. Given historic trends where the rebound effect negates a large proportion of energy savings, rapidly increasing our renewable energy supply appears to be the only way to limit carbon emission from the buildings sector.

The 2DS pathway is consistent with the optimistic assumption of reducing carbon emission from the energy system by 70% by 2060, reaching carbon neutrality by 2100. Exceeding this will require technological advancements in the generation, transmission and the storage of renewable energy.

Reference

[1] Harvey L D D 2014 Global climate-oriented building energy use scenarios *Energy Pol.* **67** 473–87

Chapter 3

Water

The most obvious common link between water and climate change is sea level rise, which will predominantly affect coastal urban areas. However, water can affect the urban environment in many ways, whether by pluvial, fluvial or environmental flooding, storm surges or by longer-term sea-level rise, the impact of large amounts of water flowing through an urban area can cause significant damage to property and infrastructure, and pose a significant risk to human life. This chapter will examine the effects of flooding events, how these may change as a result of climate change and what potential solutions can be implemented to mitigate risk.

3.1 Introduction

Over recent years, people in the UK and beyond have experienced increasing weather variability, stronger storms and more intense rainfall. There is perhaps still too little evidence to lay these events solely at the feet of anthropogenic climate change, largely due to the infrequency and variety of events and lack of historical records. Nevertheless, in recent years we have seen intense rainfall cause flooding, disrupt transport links, destroy property and infrastructure and in some cases create public health risks and loss of life. It is important to understand that how urban areas cope with water is linked not only with the amount of precipitation but also the capacity of drains and rivers, the state of the tide and atmospheric pressure. This was highlighted by the destruction of the village of Boscastle (Cornwall, UK) on 16 August 2004, which resulted from a series of factors that will be covered more generally later in this chapter. On that day, over 1000 trees were destroyed and 20-years' worth of river sediment was dumped in the village, 75 cars and six buildings were completely washed out to sea and over 100 other buildings were severely damaged, but luckily no one died.

doi:10.1088/978-0-7503-5262-8ch3

So, what caused this disaster? On the 16 August, 185 mm (7′) of rain fell on the ground above Boscastle in the space of 8-h, the peak rainfall in Boscastle itself was recorded at 89 mm (3.5′) in an hour, more than a months' worth of rainfall! The rainfall was incredibly localised, of the 10 rainfall gauges in the area (all within a few miles) half showed less than 3 mm of rain that day. The local geography is significant, land rises steeply and Boscastle lies in a valley with a natural harbour, there are over 10 springs within a mile radius of the village which all converge in the village, meaning there is always water flowing out to sea regardless of precipitation. To make matters worse, the flood coincided with high tide. Tides in August are relatively large, almost the size of spring tides, this coupled with the storm surge caused by the low-pressure system above the area, magnified the river flooding and prevented water flowing as quickly out to sea.

The torrential rain caused upstream rivers to bust their banks, causing 2 000 000 m^3 (2 billion litres) of water to flow down into the village, a full bath tub is around 1 m^3! The continuing rainfall caused the river estuary to rise by 2 m (6′6″) in 1 h. A second flash flood was caused by cars, trees and other debris becoming lodged under a bridge creating a dam, before breaking and releasing a second 3 m (10′) wave. Unlike static flooding, this was dynamic and water speed was over 4 $m\,s^{-1}$ (10 mph, 14 $km\,h^{-1}$), more than enough to cause damage to buildings and wash cars out to sea (figure 3.1). It has been estimated that more than 20 000 000 m^3 of water flowed through Boscastle that day. The cause is not entirely natural, while the rainfall amount and intensity were extreme, much of the damage was caused by anthropogenic activities. In addition to

Figure 3.1. The local shop after the flood, this was one of the few buildings left standing (credit: Ben Evans). This Boscastle 0011 image has been obtained by the author from the Wikimedia website, where it is stated to have been released into the public domain. It is included within this article on that basis. https://commons.wikimedia.org/wiki/File:Boscastle_0011.jpg

Figure 3.2. View upstream after flood waters had subsided, note the damage to the building façade (credit: Ben Evans). This Boscastle 05 image has been obtained by the author from the Wikimedia website, where it is stated to have been released into the public domain. It is included within this article on that basis. https://commons.wikimedia.org/wiki/File:Boscastle_05.jpg

the cars and other debris becoming lodged under a bridge causing a second larger flash flood, changes in farming practices upland of the village with the removal of trees, hedges and improved sewerage caused the water to flow much faster than it would have done in the past (figure 3.2).

Such extreme events will be covered further in chapter 8, but such disasters are linked to many contributing factors, many of which will be magnified as a result of climate change. The impacts for coastal or low-lying areas or those adjacent to rivers will be significant and are discussed in the rest of this chapter.

3.2 Sea level rise

Sea level rise is probably the most well-known and talked about aspect of climate change. While increasing temperatures at the poles promote melting of the polar ice caps, this is not the main cause of sea level rise. In the same way that an ice cube in a glass of drink does not alter the liquid level in the glass as it melts, floating ice caps do not contribute to sea level rise. However, land locked ice masses and glaciers will, but this contribution is small compared to the primary cause, which is thermal expansion of the oceans as they warm. Due to the volume and dynamic nature of the oceans, they have a massive thermal inertia and will continue to warm and rise for several centuries after air temperatures have stabilised. This means that while we consider climate projections to the end of this century in this book and some projections show stabilisation of air temperatures, the estimates of sea level rise are only the tip of the iceberg so to speak.

Sea level rise is a major issue for low-lying countries such as Bangladesh or Micronesia, where there is little infrastructure to hold back the sea, which could see large areas of their land mass become submerged. However, developed countries will also face significant challenges. Much of the Netherlands is already below sea level (see figure 3.3), and therefore rivers have to be managed carefully so that water flows out to sea without extensive flooding. As such, extensive infrastructure is employed to hold back the sea and drain the land. Parts of the Netherlands are up to 7 m below sea level. As sea levels rise, dykes will have to be made taller and pumps upgraded to be able to cope with greater amounts of water, to avoid extensive flooding. Due to the low-lying nature of the country, the Dutch have adopted the philosophy of living with water rather than trying to control it as in other countries, which can be costly and very energy intensive. Hence, there is much that other countries can learn about how to manage water in both urban and rural areas.

Other countries also employ measures to hold back the sea. The UK has amongst the largest tides in the world and like many other European countries its cities are built on, or alongside rivers and estuaries. The port city of London is no exception. Due to the topography along the Thames estuary the Thames barrier, completed in 1982, was created to protect the main financial centre of the city and to protect over 200 000 properties in and around London. The barrier was originally designed to protect London from very high flood levels (up to 1 in 1000-year events) and to close if the tidal water level in central London were expected to exceed 4.87 m (16′) (a 1 in 100-year event). At the time of original conception (circa 1970) it was expected to be

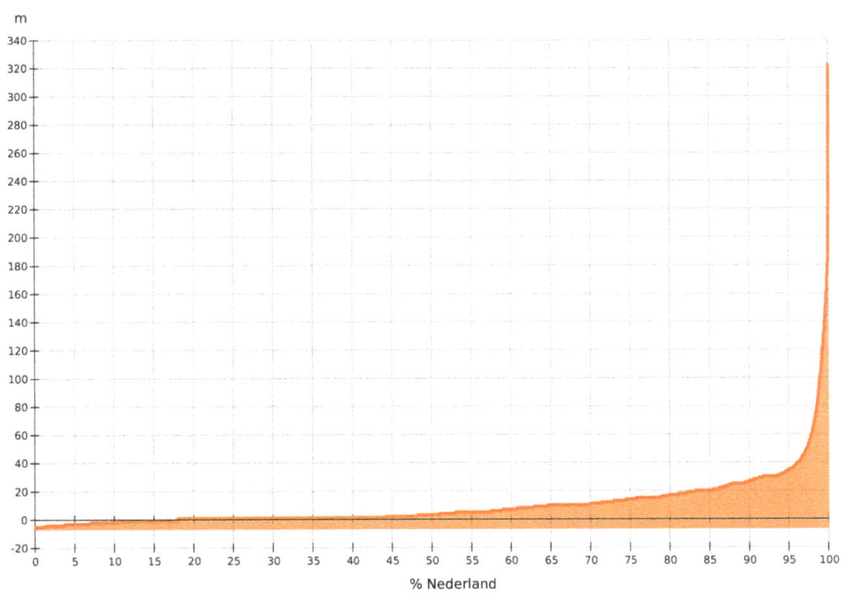

Figure 3.3. Cumulative distribution function of height above sea level for the Netherlands as percentage of land mass. This Hypsometric curve of the Netherlands image has been obtained by the author from the Wikimedia website where it was made available by User:PawełS under a CC BY-SA 3.0 licence. It is included within this article on that basis. It is attributed to User:PawełS. https://commons.wikimedia.org/wiki/File:Hypsometric_curve_of_the_Netherlands.svg

used two to three times per year with a useful lifetime of up to 2070, based upon the understanding of climate change and sea level rise at the time. Between completion in 1982 and March 2007, the barrier was raised 100 times to prevent flooding, greater than the initial estimates but still within expected limits. However, over the winter of 2012, the Thames barrier was raised 50 times, resulting in recommendations to the Government and the Environment Agency that a new barrier be considered. The Thames Estuary 2100 plan considered options for the future protection of London and its environs from tidal flood risk for the rest of this century.

The Thames Estuary 2100 project found that by upgrading and modifying the existing Thames barrier, protection could be maintained until the 2070s. The strategy was designed to be able to cope if sea levels rise was greater than the maximum stated in the IPCC 4th Assessment projections. A high-plus-plus (H++) scenario was devised to be used for such contingency planning. The H++ scenario is based upon extensive land locked ice sheet and glacier melting, in addition to thermal expansion of the oceans and hence represents a low probability, high impact scenario. The H++ scenario places an upper limit of 2.5 m sea level rise during the twenty-first century, in addition to upper estimates for storm surge (covered later), this level of sea level increase is feasible given current understandings but is considered very unlikely. Since the use of the H++ scenario in the Thames Estuary 2100 project, it has also been used for contingency planning for the provision of new sea defences at other locations around the UK.

A similar project is currently being built in Venice (Italy) to protect the historic city from rising sea levels and increasing storm surges. The MOSE project (*MOdulo Sperimentale Ellettomeccamico*) is due to be completed in 2018 and is intended to protect the city of Venice and the Venetian Lagoon from flooding. Similar to the Thames barrier, gates are raised to keep high tides and storm surges out. The low-lying nature of Venice means that large sections of the city are flooded if the defences are breached by even a small amount, as can be seen in figure 3.4. The MOSE system was deemed a better way to protect the heritage of the city than constructing sea walls, which would impact upon the cultural aspects of the historic city.

Venice experiences much smaller tides than the UK, and as such the barrier has only been designed to protect against water heights of 3 m (10′), including 60 cm (2′) of sea level rise. To date, the highest tide (and storm surge) was 1.94 m (6′4″). Hence, the MOSE project might not be as resilient to future climate change as the Thames barrier; however, it covers a much larger area and spans some 18 km.

The UK climate projections (UKCP09 & UKCP18) provide estimates of possible sea level rise over this century, along with estimates of extreme sea levels due to storm surge (skew surge). Figure 3.5 shows the range of likely sea level rise for the UK. As for other climatic variables, the projections of sea level rise are probabilistic, as represented by their relative cumulative distribution function frequency (i.e. the 5% level represents the value where 5% of projections are less than this value). While such information is not particularly easy to interpret at first glance, the probabilistic projections from UKCP09 allow a greater consideration of the risk of climate change.

Figure 3.4. Flooding of the Piazza San Marco Venice in 2012. This San marco aqua alta image has been obtained by the author from the Wikimedia website where it was made available by User:Abxbay under a CC BY-SA 3.0 licence. It is included within this article on that basis. It is attributed to User:Abxbay. https://commons.wikimedia.org/wiki/File:San_marco_aqua_alta.JPG

3.3 Storm surge

The height of our seas and oceans is not static, it varies not only with the seasons and cycles of the Moon affecting the heights of our tides but also with the weather! A skew surge or storm surge is the height difference between a predicted astronomical high tide and the observed high tide. Storm surges occur, as the name suggests, during storms or when a low atmospheric pressure weather system is above the ocean. High pressure weather systems push down on the ocean's surface, causing the water level to rise elsewhere there are low-pressure systems. This can result in higher than expected high tides if the astronomical high tide corresponds with a low atmospheric pressure event. The lower the pressure, the higher the water rises. As you might have guessed, the lowest pressure systems are often associated with storms often bringing intense rainfall and high winds. This can create an additional problem in that the higher than normal tide can prevent river water and surface runoff from intense rainfall from flowing out to sea, as was the case in Boscastle in 2004.

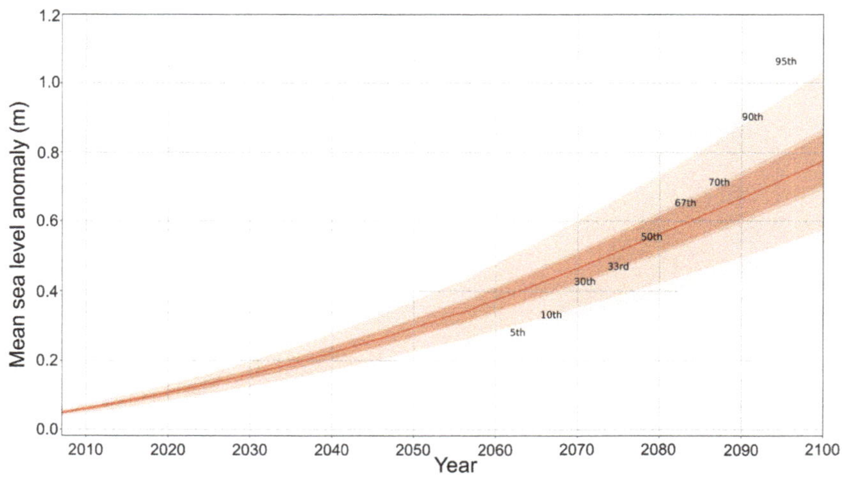

Figure 3.5. Estimates of sea level rise over time for different (cumulative distribution function) probability levels. Data for Avonmouth, in the Bristol Channel, (51.5°, −2.75°) using RCP8.5 and a baseline of 1981–2000 (data: UKCP18 under Open Government Licence v3.0).

Storm surges can lead to extensive flooding and are dangerous for people living in low-lying coastal areas. Sea levels rise roughly 1 cm for every millibar (mbar) change in pressure, this can be further increased by the storm approaches the coast, the shape of the coastline and the seabed. For example, in 2005 when Hurricane Katrina approached the US coast, storm surges in excess of 8 m were generated in some areas, causing widespread damage to the city of New Orleans where the sea defences could not cope with the water level. Storm surges for the UK are typically more modest, ranging between 1 and 3 m for different UK locations for a 1 in 50-year type event. Return periods such as this refer to the average amount of time between similar events. However, it is a mistake to think that if a 1 in 10-year event occurred this year we would not see a similar event for another 10-years. A 1 in 10-year event could occur twice in the same year or not at all for several decades. It is best to think of these return periods as the probability of an event occurring in any given year, regardless of when the last similar event was (i.e. 1 in 10 years would be 10%, 1 in 100-years, 1% and so on).

As the climate changes, storms are expected to become more intense and the return periods of different events are expected to change, which will be covered more in chapter 7 – Weather Extremes. Additionally, reduction in the density of seawater due to thermal expansion (sea level rise) will lead to a relative increase in storm surge height over time. Figure 3.6 shows how the height of a storm surge associated with a given return period of storm may change over time in mm/year, while figure 3.7 shows the absolute height (tidal + skew surge) of the sea for a greater range of return periods. Both plots are for Avonmouth in the Bristol Channel (UK), which experiences some of the largest tides in the world. The UKCP18 data allows a greater exploration of extreme events than was possible with the UKCP09 data.

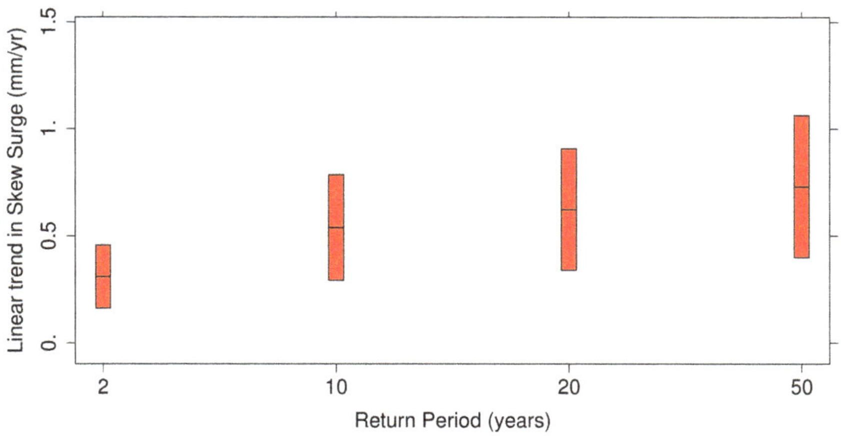

Figure 3.6. Expected change in storm/skew surge height per year for different return periods. The boxes define the 5th, 50th and 95th percentiles for Avonmouth in the Bristol Channel (51.5°, −2.75°) under the A1B emissions scenario (data: UKCP09 under Open Government Licence v3.0).

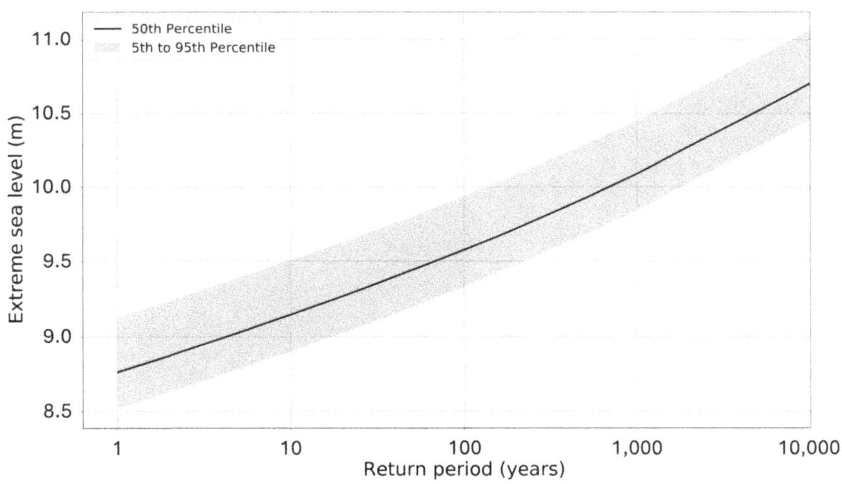

Figure 3.7. Plume plot of estimated absolute extreme sea levels (tide and skew surge) by 2100 for different return periods. The plot shows 5th, 50th and 95th percentiles for Avonmouth in the Bristol Channel (51.5°, −2.75°) under RCP8.5 (data: UKCP18 under Open Government Licence v3.0).

3.4 Flooding

Floods can have many causes. Non-coastal flooding can be split into three main types: fluvial (rivers), pluvial (surface flooding) and environmental (groundwater flooding). The most devastating floods are typically as a result of rivers bursting their banks, particularly in urban areas where the impact on neighbouring property and infrastructure is immediate. However, periods of intense rainfall can cause surface water flooding and possible 'flash floods'. In this case, rainwater has not had time to be absorbed into the ground (percolation), either because of the surface type, many urban surfaces are impermeable to water or because the available drainage, whether ditches or sewers, are unable to cope with the volume of water. If the geology of the ground is permeable to water, then it is possible for flood water to come out of the ground if the water table raises high enough. Groundwater flooding can be particularly damaging because it can be sustained for several months. This happens because the water simply does not have anywhere to go, rivers and lakes will likely already be at capacity and the ground is completely saturated.

The effects of climate change on rainfall are complex—a warmer Earth will lead to greater evaporation of water from the oceans which will lead to more rainfall. However, changes to global surface temperatures will affect global convection currents and may well alter the path that weather systems travel across the Earth (as discussed in chapter 1). The UK current estimates of climate change show very little deviation in the amount of annual rainfall under any of the emissions scenarios (figure 3.8), despite this there will be a shift in when rainfall occurs.

Currently, the UK's climate is defined under the Köppen Geiger climate classification as CFB (temperate marine), implying we have a mild climate with no distinct dry or wet season. Climate change will see a shift in rainfall from summer to winter, with the amount ranging from effectively no shift at the lowest estimates of climate change under UKCP09 up to ~70% shift in rainfall amount from summer to winter for the highest estimates. This poses particular risks for urban areas because the intensity of winter rainfall is expected to increase not just the total amount. Figure 3.9 shows maps of the average daily winter precipitation and how this may vary over time. As you might expect, the majority of rainfall falls on high

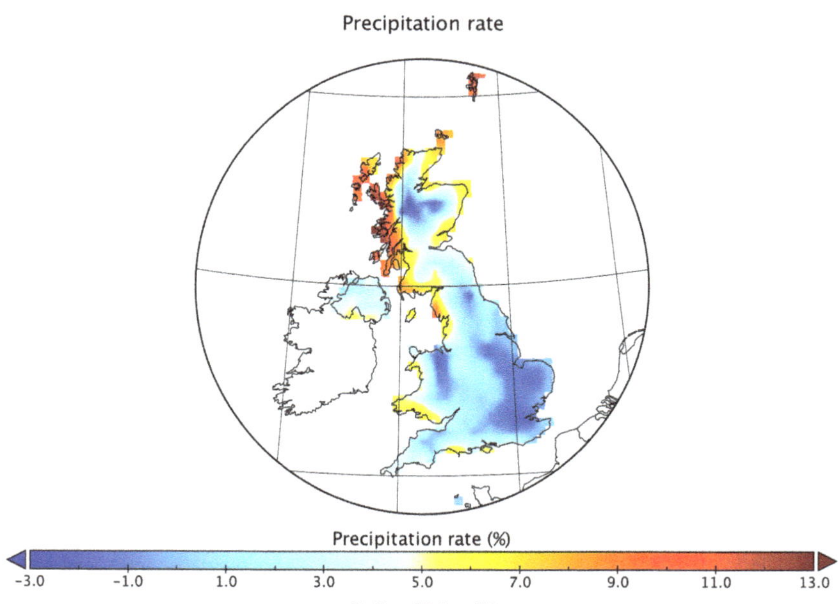

Figure 3.8. Relative change in annual precipitation by the end of the century (2080s) (data: UKCP18 RCP8.5 50th percentile under Open Government Licence v3.0).

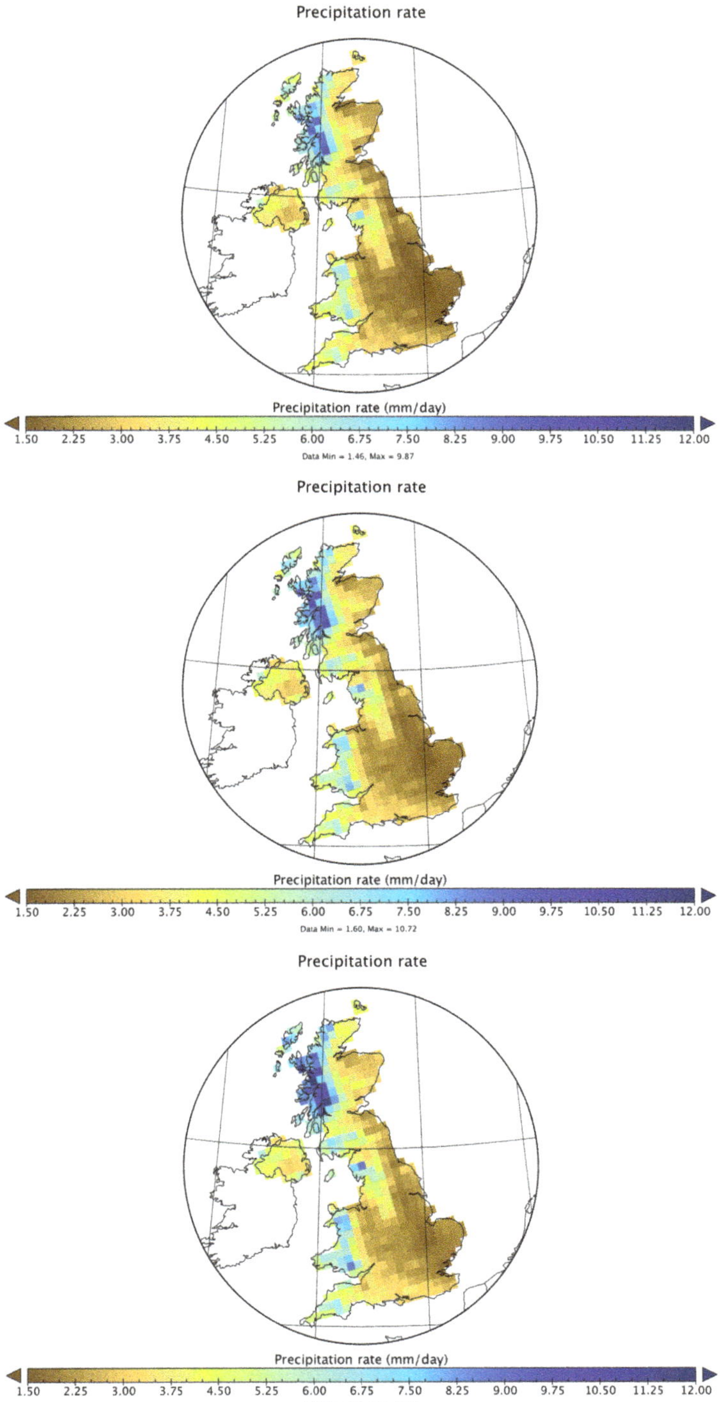

Figure 3.9. Average daily winter precipitation, central estimate for the 2020s (top panel), 2050s (middle panel) and 2080s (bottom panel) (data: UKCP09 A1FI 50th percentile under Open Government Licence v3.0).

ground, but also predominantly along the south and west coasts. Urban areas in the south and west will have to adapt to cope with increased volumes of rainfall because current infrastructure may not be adequate.

Many historic cities are located near rivers or estuaries because the waterway provided both food and transport, allowing trade with other cities and countries. This required these waterways to be tamed, their routes fixed, no long able to change according to geography and geology over time. While this allows a river to become a part of a city's infrastructure and a local amenity, there are drawbacks. As the city grows, development encroaches upon land that would have previously been allowed to flood. In addition, the riverbanks are sometimes reinforced with stone gabions, concrete or sheet steel piling to prevent the river changing course. Although this has the desired effect of confining the river to a set course, it also confines the capacity of the river to a set volume of water, which if exceeded can cause considerably flooding. This has led to the construction of flood relief channels in some cities to provide additional capacity when the river exceeds a certain height, to protect property and infrastructure. Exeter is one such city, where a flood relief channel (figure 3.10) provides additional capacity, preventing the river Exe from flooding the nearby mainline train track and nearby roads in the St David's area of the city.

Like other rivers, the Exe has its source on high ground, in this case Exmoor in the north of Devon. The photos shown in figure 3.10 show the river Exe's flood relief channel in the summer of 2016 (top panel) showing its typical water level and again in late autumn (bottom panel), this time with an elevated water level, covering the footpaths and part of the grass banks. You can see the spillway from the river at the end of the flood channel with the river beyond, normally water is not flowing over this spillway as shown in the top photo. The flood relief channel is approximately 3–4 m below the height of the river at this point (to the right of the bank in the photo) and rejoins the river approximately 1 mile downstream (1.6 km). You may notice from the photo that it is not raining, the only recent rainfall occurred on the 19 November and Exeter itself received minimal precipitation, located as it is in the rain shadow of Dartmoor. Exmoor, however, received approximately 25 mm (1′) of rainfall on that day. This highlights an important issue with urban rivers, it is not just the rainfall falling on the urban area and rapidly flowing into a river that cause flooding but there can also be delayed flooding from rainfall that occurred hours to days earlier upstream. Figure 3.11 is a photo of the river Exe adjacent to the flood relief channel taken on 20 November 2016. There is approximately 30 cm (1′) clearance between the river and the road bridge at this point in time.

So long as the flood relief channel is able to take any excess water, this level will not rise further and the bridge is protected. However, this is dependent upon the flood relief channel having adequate capacity, not just in terms of total volume but also how much water can flow in and out per second. At the time of writing, the flood defences along this flood relief channel and further down the river are being raised to provide increased capacity and continued protection in the future.

The flood relief channel featured here serves its purpose but has to rejoin the main watercourse at some point. Ideally, this would be outside of the urban area, but often this is not the case due to either geography or existing developments. In the case of

Figure 3.10. River Exe flood relief channel 15 August 2016 (top) and 20 November 2016 following a day's heavy rainfall on Exmoor (bottom).

the river Exe, the flood relief channel ends after approximately 1 mile (1.6 km) at the site of an old water mill (which is now a pub). On one side behind the pub the bank rises steeply to the road and the city beyond, on the other side of the bank there is a grass bank followed by flat playing fields with houses beyond. The following two

Figure 3.11. Road bridge over the river Exe on the 20 November 2016.

photos (figure 3.12) were taken from these playing fields showing how the river floods after a more prolonged period of heavy rainfall, in this case the winter of 2012. As we can see from figure 3.12, the river at this point has risen significantly (>2 m) and the river basin (including footpaths and cycleways) is at capacity. This area had been designed to cope with such volumes if water and the level of protection is currently being increased; however, more historic areas such as Exeter Quay a little further downstream, cannot cope as well. While recent apartment developments are raised above the river, the historic buildings (which are now shops and restaurants and tourist attractions like the old customs house) are low lying and are only ~1 m above the standard river height. The volume of flood water which passed through the Quay during the winter of 2012 caused significant damage to both property and peoples' livelihoods. In this respect, for Exeter, like many cities, it is the historic areas that are most vulnerable to flooding because they were not designed to cope with such large volumes of water, which are the result of upstream developments, increasing peak water flow rates. The question then becomes how can we both protect and maintain the heritage of our cities?

3.5 Flash flooding

Flash flooding can occur for several reasons, both natural and anthropogenic. Flash flooding is different from normal flooding, which might be caused by prolonged rainfall, saturating the ground and raising the water table to the extent that water does not soak in or drain away. Instead, flash flooding is most common after periods of intense rainfall, particularly when following periods of warm dry weather, so the

Figure 3.12. The River Exe at the confluence of the river (right-hand side) and flood relief channel (left-hand side), as it looks normally (top panel) and during the floods of 2012 (bottom panel). Image credit: James Webb

ground becomes hard, compacted and impermeable. The state of the ground is important because it is highly unlikely that enough rainfall will fall in a given location to cause flooding damage, instead water is transported to, and converges in locations susceptible to flooding. This can be over the surface of the ground, either because of previous weather conditions or because of the presence of man-made impervious materials, such as roads.

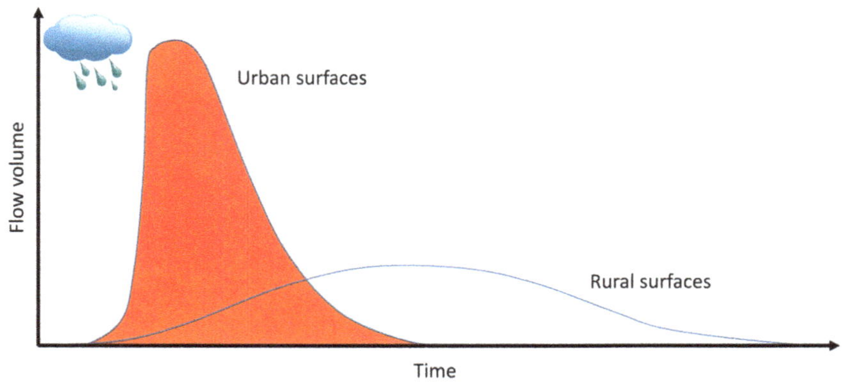

Figure 3.13. Illustrative hydrograph for typical urban and rural surfaces.

Figure 3.13 shows an indicative hydrograph of the typical response for rain falling on an urban surface rather than a rural one. The area under each of the curves is the same because the same volume of water is involved, but more water flows for a shorter amount of time over the hard, impermeable surfaces associated with an urban area.

Surprisingly, it is not just urban development that can increase peak flow rates and cause flash flooding. The events of 2004 in Boscastle (as discussed earlier in this chapter) were attributed not only to intense rainfall but also clearing of scrubland uphill from the village. In this way, agricultural practices can also lead to increased flood risk, clearing of land, removal of hedges, shrubs, trees etc can all increase peak water flow rates. All vegetation, whether trees, shrubs of even grass slows water down, through interception of falling rain and increased surface roughness. There is also the added benefit that rainfall intercepted by a plant canopy may never reach the ground, instead it may be held long enough to evaporate directly from the plant canopy. Plants also use water from the ground and this is passed as water vapour out through stomata in leaves in a process called transpiration. This helps prevent water saturation of the ground and when coupled with slower moving water the likelihood of the water being absorbed into the ground through infiltration (also known as percolation) is increased. It is clear then that vegetation or land surface types that slow water down and give it time to be absorbed or evaporated are vital not only to help prevent flash flooding but also help prevent pluvial and environmental flooding.

Climate change will increase not only shift year-round rainfall to predominantly winter rainfall but will also increase the intensity of rainfall events. Figure 3.14 shows how the amount of rainfall on the wettest day in winter will change by the end of the century under an upper estimate of climate change, while figure 3.15 shows the spatial distribution of rainfall for a 1 in 100-year event. High ground already receives a large amount of rainfall. For instance, according to the underlying data, parts of Western Scotland received >360 mm over the five-day period illustrated in figure 3.15. Thankfully, this will not significantly increase. However, lower lying areas with large urban areas such as the south and east UK will see a large increase in intense rainfall (see figure 3.14). For example, the south coast around Portsmouth

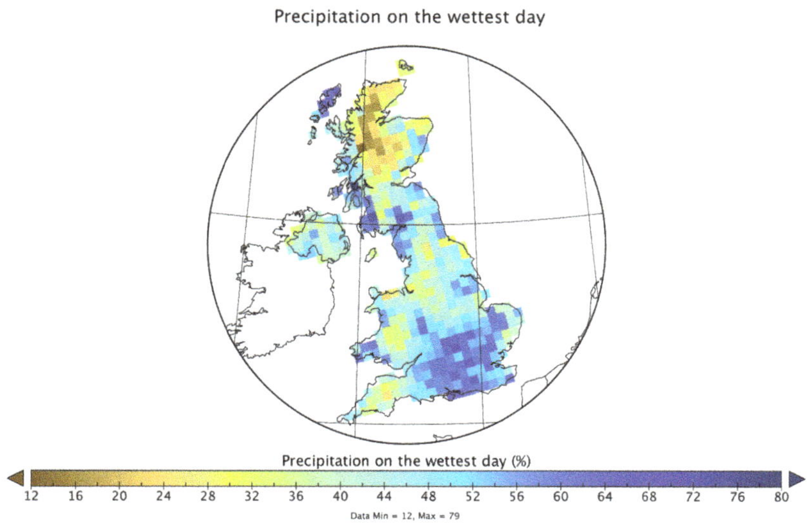

Figure 3.14. Change in the amount of precipitation on the wettest day in winter by the end of the century (data: UKCP09, A1FI 90th percentile under Open Government Licence v3.0).

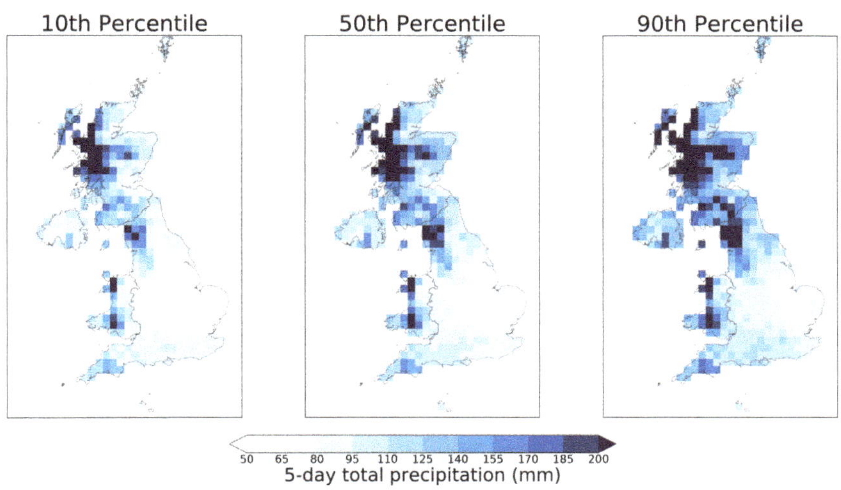

Figure 3.15. Winter 5-day total precipitation by 2050 for a 100-year return period event (data: UKCP18, RCP8.5 under Open Government Licence v3.0).

and Chichester, which is already susceptible to flooding, may well see a 79% increase in the amount of rainfall to fall on the wettest day in winter. Such intense rainfall will likely be accompanied by a large storm/skew surge, which will amplify flooding risk for these coastal cities.

3.6 Potential solutions

Potential solutions to prevent or limit flooding in urban areas can be split into three different types, those that control water where it falls (source control), as it travels to

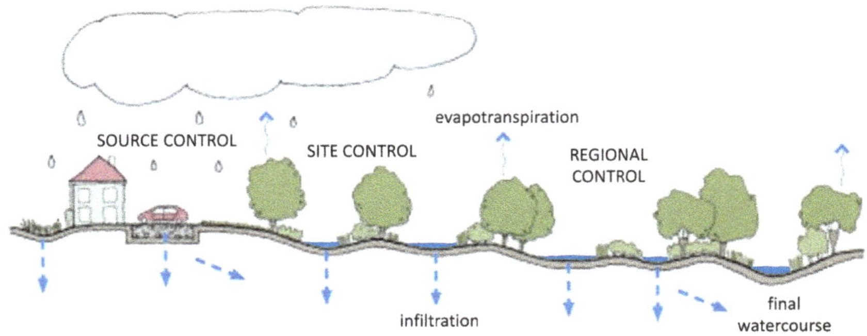

Figure 3.16. The urban water management train (adapted with permission from SUSDRAIN and CIRIA).

collection points (site control) and wider regional control. The hierarchy of these different solutions is referred to as the urban water management train, as shown in figure 3.16.

3.6.1 Source control

Controlling water where it falls is the first step in limiting flood risk, this typically involves slowing water down so that that it has time to be absorbed into the ground. For most new developments, this takes the form of a sustainable (urban) drainage system (SuDS) plan. SuDS is a sequence of water management practices spanning both source and site control that aim to drain surface water in a more sustainable manner than simply routing water runoff to drains and through a pipe to a watercourse. A SuDS plan can include:

- Rainwater harvesting
- Green roofs
- Permeable surfaces
- Filter and infiltration trenches
- Swales
- Detention basins
- Bioretention systems or raingardens
- Retention basins, wetlands and ponds

Collecting water from both a roof (rainwater harvesting) and green roofs such as the one shown in figure 3.17 slow rainwater down and prevent it from flowing directly to a drain. In the case of rainwater harvesting, the water is collected in a tank and used for non-potable uses around the home, such as flushing toilets, washing clothes or watering the garden. Water can be stored for days or weeks before it is used and enters drains or the ground. Although green roofs only store water in the soil substrate, this water is used by the plants and will have a net cooling effect on the roof and the surrounding air. These aspects will be covered later in chapter 7, in terms of controlling water, a green roof's primary purpose is to reduce water flow rates, both through collection and use by the plants. In addition, the

Figure 3.17. A sedum green roof atop the British Horse Society headquarters. This British Horse Society Head Quarters and Green Roof.jpg image has been obtained by the author(s) from the Wikimedia website where it was made available by Sky Garden Ltd under a CC BY-SA 4.0 licence. It is included within this article on that basis. It is attributed to Sky Garden Ltd. https://commons.wikimedia.org/wiki/File:British_Horse_Society_Head_Quarters_and_Green_Roof.jpg

Figure 3.18. Examples of permeable paving (reprinted with permission from SUSDRAIN and CIRIA).

rough surface presents a smaller steadier flow of water to a drain than a tiled roof (for example).

For other typically non-permeable areas such a car parks, pathways and pavements, we can make the surface rougher to slow water down and use permeable materials and constructions to encourage percolation of water into the ground. Such permeable surfaces are now fairly typical for out-of-town developments such as supermarket car parks which have been built on an area that was historically flood plain. Figure 3.18 shows two examples of permeable paving suitable for car parks and pathways; however, other materials can be used such as resin bound aggregate which is porous to water. The paving shown in figure 3.18 presents a rougher surface to the water than (for example) tarmac, and hence it will flow slower over the surface. This relieves stress on drains, which have a finite capacity, and provides more time for water to pass through gaps between the pavers and to percolate into the soil below.

Such permeable surfaces can be coupled with a water storage tank, either as part of a rainwater harvesting system or just as an attenuation tank, providing storage before water flows to a watercourse or drain. Figure 3.19 shows an example of how this could be implemented, the paved road or carpark consists of pavers, a gravel substrate and a water permeable membrane resting upon plastic attenuation crates. The storage tank is made up of modular attenuation crates (figure 3.20) surrounded by a water proof membrane. The crates are available with different loading capacities and can be used for purposes ranging from under a domestic patio or sports pitch to a lorry loading bay. The modularity of the attenuation crates allows great flexibility in the shape and size of the water collection area and also volumetric capacity of such a system.

An alternative source or site control methodology to providing a permeable surface for water to pass through is to use conventional impermeable constructions and to guide water to an area where it can collect and drain away. A bioretention system or a raingarden such as the one shown in figure 3.21 is a shallow hole or depression containing vegetation that collects rainwater runoff from surrounding impermeable surfaces, such as roofs, driveways, roads and pavements, providing the opportunity for the water to percolate into the ground.

Raingardens reduce surface runoff by allowing water to collect and be absorbed into the ground rather than flow into drains or across the surface. These can range from large areas such as the one shown in figure 3.21 to collect rainfall from a nearby road and pavement, to smaller areas collecting rainfall from a single roof or patio,

Figure 3.19. Example of permeable paving coupled with water storage.

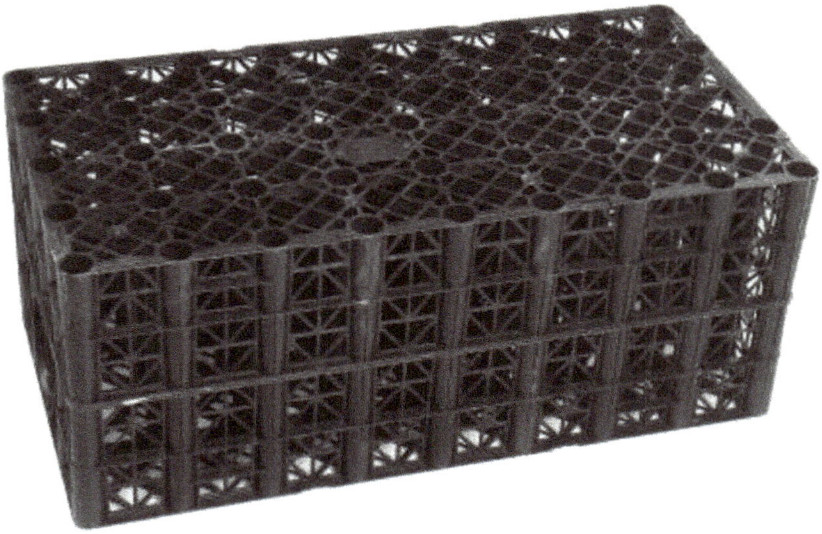

Figure 3.20. Image of a water attenuation crate for stormwater runoff (reprinted with permission from SUSDRAIN and CIRIA).

Figure 3.21. A raingarden in Portland Oregon, designed to capture surface water from the road and pavement. Image courtesy of the United State Environmental Protection Agency.

replacing a soakaway. Raingardens also have the benefit of improving the water quality of nearby waterbodies by preventing pollution flowing into them and increasing groundwater for plants rather than sending stormwater straight out to sea.

3.6.2 Site control

Site control is an extension of the principles discussed above for source control, just on a slightly larger scale. This typically involves storing water runoff from a

Figure 3.22. Swale drainage ditch.

development or as an attenuation strategy at the culmination of a SuDS plan. Examples of site control include drainage ditches and swales (figure 3.22), which are simply depressions in the ground for water to collect and to drain away or be guided towards a suitable watercourse or drain. These have limited capacity and are typically used to collect water from a road or pathway.

On a larger scale, detention and retention basins provide a larger volume of water to collect and to soak away, and can harbour many different species of flora and fauna, being a source of biodiversity for the site. The primary difference between a detention basin and a retention basin is the presence of water. A detention basin (figure 3.23) simply holds water temporarily after a rainfall event and the water drains away over time—the basin is typically empty. A retention basin (figure 3.24), on the other hand, as the name suggests, retains water and could be considered a pond or reservoir with excess capacity.

3.6.3 Regional control

Regional control typically refers to the management of watercourses and the management of flood plains to protect urban areas. One such example is the Somerset Levels in the southwest of the UK, which contains several watercourses some natural and others man-made with sluices to control water flow across the levels and out into the Bristol Channel. King's Sedgemoor Drain (hereafter Drain) is one such man-made watercourse, which diverts the River Cary in Somerset to discharge into the River Parrett. As the name suggests, this watercourse was created between 1791 and 1795 to drain the local King's Sedgemoor peat moors, which were generally flooded with the water only able to leave via evaporation. This was due to the existing drainage outfalls in to the River Parrett being too high to be of much use. The 10.5 mile (17 km) channel of the Drain was completed in 1795, resulting in improved drainage to the surrounding area of 11 000 acres (4500 ha). The Drain has been upgraded over the last few centuries, being widened and deepened in

Figure 3.23. A detention basin designed to capture rainfall from a nearby development.

Figure 3.24. A retention basin, capturing and storing water from a nearby development. This Trounce Pond.jpg image has been obtained by the author(s) from the Wikimedia website where it was made available by User:Drm310 under a CC BY-SA 4.0 licence. It is included within this article on that basis. It is attributed to User:Drm310. https://en.wikipedia.org/wiki/Retention_basin#/media/File:Trounce_Pond.jpg

places with the addition of tilting sluice gates to allow more flexible water level management.

Flood waters are removed from drainage ditches on the Somerset Levels by pumping stations, which were originally steam powered but are now electric. The Drain itself operates entirely by gravity, this means that it is dependent upon the state of the tide in the Bristol Channel, which has one of the largest tidal ranges in the world. There were plans in 2002 to add a pumping station to the Drain, which would have allowed water to be pumped from the Drain into the estuary at all states of the tide, but these were not followed through. However, the presence of the Drain and its interface with other rivers such as the Sowy, Parrett and Cary, along with numerous ditches, rhynes and associated infrastructure, allows water to be managed on the Somerset Levels, including the selective flooding of certain areas, in order to protect the urban centres of Bridgewater and Taunton.

However, such extensive water infrastructure is not always enough, particularly in this case where a large collection area is dependent on the state of the tide to drain efficiently. The winter of 2013–14 is one such example, which saw the Somerset Levels flooded for several months. The combination of prolonged rainfall, large tides and a storm surge meant that the rivers and drains were at capacity, and water was only able to flow out to the estuary for a few hours each day before sluice gates had to be shut again to prevent tidal flooding. The prolonged rainfall saturated the ground, leading to environmental flooding across the Levels, where water simply rises out of the ground in low-lying areas rather than having to have flowed there, bypassing any flood defences. Figure 3.25 shows the result of the extensive rainfall

Figure 3.25. Photo showing the confluence of the rivers Parrett and Tone during the flooding of February 2014.This Confluence of the Parrett and Tone at Burrowbridge during flooding in February 2014.jpg image has been obtained by the author(s) from the Wikimedia website where it was made available by User:Rodw under a CC BY-SA 3.0 licence. It is included within this article on that basis. It is attributed to User:Rodw. https://commons.wikimedia.org/wiki/File:Confluence_of_the_Parrett_and_Tone_at_Burrowbridge_during_flooding_in_February_2014.jpg

over the winter and the widespread flood across the Somerset Levels. In this photograph, the two rivers Tone and Parrett are at capacity where they meet. However, the dykes are still visible above the rivers and flood waters. The flooding visible in the bottom left-hand of the photograph is as a result of environmental flooding rather than as a result of the rivers having burst their banks.

3.7 Conclusions

Given the likely changes to rainfall patterns across the UK as a result of climate change, flooding is going to become a more common occurrence. The integration of a SuDS plan is a valuable tool in combating flooding. Softening the landscape, adding permeable surfaces, green roofs, water butts or rainwater harvesting, soak-aways and raingardens will all reduce the risk of flash flooding, and by extension reduce the risk of pluvial and fluvial flooding.

However, such flood reducing strategies can be negated by both heritage and geography. If the urban landscape in question is historically significant, then altering it to incorporate a SuDS strategy, or even the widening and deepening of waterways or the creation of flood relief channels, is not possible. In the same way, if the urban area is surrounded by hills, such as Boscastle, then the risk of flash flooding is greater because water will flow rapidly downhill and a SuDS plan may have little effect due to the volume and speed of the water. Instead, in this situation maintaining hedgerows, copses of trees and areas of scrubland on the hillside will slow water down before it can reach the urban area. This issue is exacerbated further if the urban area in question is coastal and relatively low lying, such as the town of Fowey (figure 3.26).

Fowey is located on the south coast of Cornwall (UK) and has an undulating topography. This means that rainwater will flow quickly into the valleys and streams (and the drains with Fowey itself) and other low-lying areas. The Environment Agency estimate that there are 290 properties, roads, a wastewater treatment works and two electricity substations at risk from a 1% probability (1 in 100 year) flood in Fowey. Over this century, it is estimated that the number of properties at risk will increase by around 14. The Environment Agency also estimate that there will be 20% more peak flow in watercourses based upon predictions of climate change, increased urbanisation and changes in land use. This increases the likelihood of

Figure 3.26. Panoramic view of the Fowey Estuary, which is home to many yachts and pleasure craft during the summer months.

large-scale flooding in and around Fowey. Flooding of infrastructure such as the water treatment works and the two substations can have disastrous effects and will likely be subject to increased flooding defences. This may not seem like a major issue but for a town like Fowey much of the town's income is from tourism based on the town's location and aesthetic.

Fowey is a quaint historic Cornish seaside town, with narrow streets, which due to the surrounding topography have a steep incline (figure 3.27) down to the harbour and the main tourist centre. In the historic part of the town, the streets are narrow, with no pavement and are just wide enough for a car and a pedestrian. The streets are lined by stone walls, further confining where water can and can't go.

The landscape levels off as you reach the town centre (Fore Street), which together with the promenade and marina constitute the main tourist centre of the town. Fore street runs parallel with the estuary with buildings backing onto the water and there are several openings between buildings to allow access to jetties. There are currently minimal sea defences and Fore street is less than half a metre (20′) above the average high tide level (figure 3.28).

The low-lying nature of Fore street and the main tourist centre of Fowey means that flooding is inevitable each time there is a particularly high tide, whether astronomical or due to storm surge. This also means that sea level rise is a particularly pertinent issue for the residents of Fowey, the median estimate of sea level rise by the end of the century is approximately 0.5 m (figure 3.5), which is sufficient for water to overcome the current sea wall with almost daily regularity. Figure 3.29 shows Fore Street on a sunny summer's day full of tourists and also during a spring high tide, with sand bags outside shop doors to help limit flooding.

As the events at Boscastle in August 2004 showed us, storm surges and heavy rainfall can coincide with disastrous consequences. While an entire month's rain falling in such a short period of time is certainly extreme, such events may become more common in the future. Furthermore, the coincidence of these two events is

Figure 3.27. Three streets in Fowey, note the steepness and width of the streets, water has nowhere to go but downhill.

Figure 3.28. View of Fowey sea front and harbour wall from the water, you can see the mean high tide level on the walls.

Figure 3.29. Images of Fore Street when flooded by spring tides (March 2008) and as it appears normally (July 2016). Flooding images courtesy of Fowey Renewable Energy Enterprise.

quite likely since a low-pressure system capable of creating a large storm surge typically produces a lot of rain. This can exacerbate these individual climate impacts because the rainwater within rivers and streams will not be able to flow out to sea as readily due to the storm surge. Within Fowey, there are implications for storm drains, particularly the low-lying area around Fore Street, which already suffers during storm surge events. Water can enter drains uphill but can only exit again at the bottom of the hill due to increased resistance from the storm surge preventing this water exiting into the estuary.

For towns such as Fowey, adaptation to cope with the combination of sea level rise, greater storm surges and more intense rainfall is unappealing. The H++ scenario used by the Environment Agency for a sea defences such as the Thames Barrier sets an upper limit on sea level rise of 2.5 m by 2100. This conservative scenario is likely to be used for sea defences around critical infrastructure, such as electricity substations and water treatment works, to protect them from sea level rise and increased storm surges. Such visually intrusive flood defences will impair the local aesthetic and will have an impact on tourism. While increasing temperatures may well increase tourism numbers in the shorter term (sea levels rise slower than temperatures), a long-term strategy to protect or even move the town centre will need to be employed. At the very least, openings from Fore Street to the estuary such as the one shown in figure 3.30 will need to be protected to prevent occurrences such as the one shown in figure 3.31.

Figure 3.30. Photo showing an opening from Fore Street to moorings and jetties on the estuary. Currently, there are no sea defences and water is free to flow through to the town during particularly high tides.

Figure 3.31. Photo showing the same area as figure 3.30 during a high spring tide (March 2008). Flooding images courtesy of Fowey Renewable Energy Enterprise

Fowey is not alone in this respect and this issue is not limited to small coastal villages and towns, many of our historic cities are built on rivers and tidal estuaries. Such flooding events are the entire reason for flood defence infrastructure, such as the MOSE project and the Thames Barrier, and we are likely to see other similar defences deployed to protect our cities and cultural heritage sites in the near future.

Chapter 4

Temperatures

When people think of climate change or 'Global Warming' the most obvious impact is an increase in temperatures. For somewhere with a temperate climate such as the UK where building energy use is still predominantly heating rather than cooling related, an increase in temperatures may seem like a good thing. This chapter will examine the extent of temperature changes along with what this means for human thermal comfort, productivity and the risk to life from heat stress. Finally, by looking at architecture from around the world, we examine what potential solutions can be implemented to mitigate risk.

4.1 Introduction

People often talk about global warming when referring to climate change. However, the term global warming is avoided by climate scientists for the reason that not all places on Earth will get warmer, some may get cooler as a result of changing global circulation patterns. For instance, if the Gulf stream were to move, the UK would experience a very different climate, being on a similar latitude to Moscow (Russia) and Alberta (Canada). That said, for the vast majority of urban areas it is highly likely that temperatures will increase as a result of climate change.

This poses a problem because the way we design our buildings and urban areas, and even our way of life, is determined by the background climatic conditions, particularly air temperatures. Buildings have two main purposes—to provide shelter form the elements and to provide thermal comfort. The latter being determined by the human metabolism and the local environmental conditions. Our bodies produce heat as a by-product of the biological processes going on in our bodies. As warm-blooded creatures, we maintain a core body temperature, meaning that sometimes we have to lose heat or keep heat in depending on the local environment. Generally, the easier it is for our bodies to maintain this core temperature, the more

doi:10.1088/978-0-7503-5262-8ch4

comfortable we feel. This has fed into how we design our buildings. In the UK and other northern countries, buildings are predominantly designed to keep heat in; while in the tropics, buildings are designed to keep heat out. As the climate changes, our buildings and the way we use them will have to adapt to continue to provide both shelter from the elements and thermal comfort.

The press is full of estimates of changing temperatures due to climate change, such as a 2° increase or a +4 °C world. However, these figures are a global average, the oceans' ability to store large amounts of heat means that they warm more slowly than land masses, so a global 2 °C increase is closer to 4 °C–5 °C in London (UK). In addition, some areas will warm faster than others due to latitude, altitude or proximity to the sea (for example). Figure 4.1 shows the output of a global climate model (GCM) that was used for the IPCC Fifth Assessment Report. The figure shows average near surface air temperatures for the month of June for 2017 and 2097. The scale is in Kelvin, so subtract 272 to get °C.

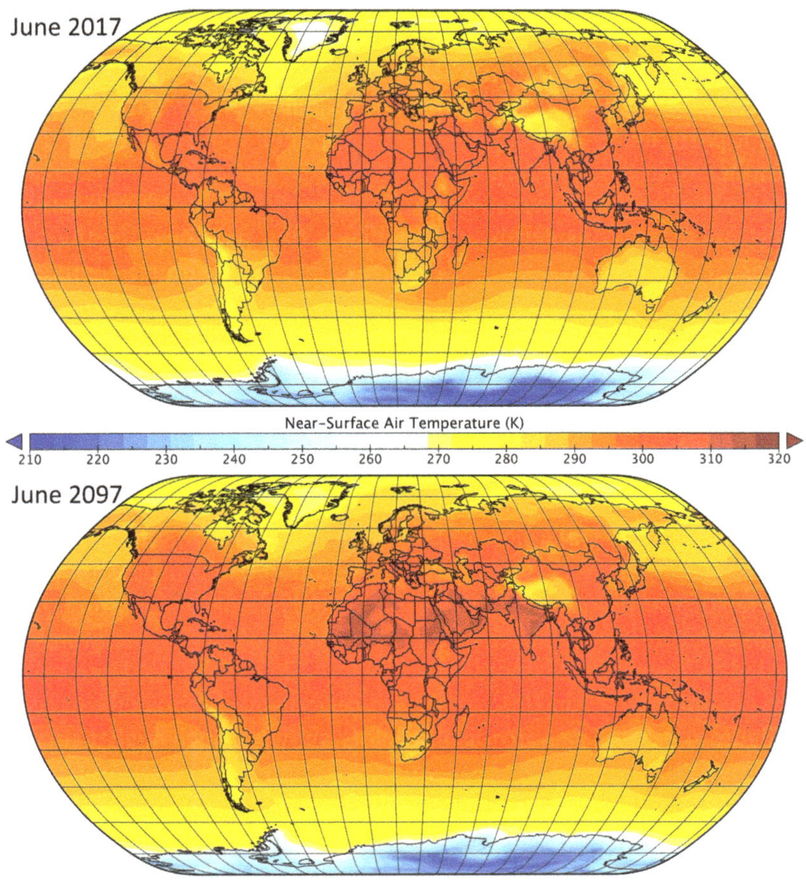

Figure 4.1. Images of near surface mean air temperature for the current climate and the end of the century under a high emissions pathway (data: HadGem2 RCP8.5 under Open Government Licence v3.0).

We can see from figure 4.1 that over the 80-year period spanning the two images that the Earth has gotten warmer, but that that the warming is not uniform, either due to topography (e.g. Himalayas) or albedo (Antarctica). While RCP8.5 is a higher estimate of climate change (see chapter 2), the consequences of this level of climate change are severe. The temperatures reached in areas of Africa and Asia (for example) are high enough to cause serious heat-related illness to human beings. If we look in more detail at the UK using regional climate model (RCM) data (figure 4.2), we can see how temperatures are projected to change over the course of this century. We can see that mean annual air temperatures in the UK can expect roughly a 4 °C increase between the 2020s (2010–39) and the 2080s (2070–99).

Despite these changes to mean temperatures across the globe and in the UK, at the lower end minimum temperatures are not expected to change dramatically. Figure 4.3 shows how the temperature of the coldest night in winter may change by the end of the century (2080s). Most areas see little rise in minimum temperatures, particularly coastal areas where temperatures are moderated by the sea. This means that climate change will have little impact on the way we specify heating systems because the peak load temperature used to determine the capacity of say a boiler will not change much. This implies we may need to find a different way of heating our homes in order to maintain efficiency. Boilers are typically inefficient if only used at part of their overall capacity. In order to be efficient and yet meet the peak load demands of a particularly cold night, multiple smaller systems may need to become the norm rather than a single large boiler.

The small change in minimum temperatures implies that the maximum temperatures must increase considerably to achieve the predicted changes in mean temperatures. The distributions of daily, seasonal and annual temperatures are not just shifted, they are stretched as well. Figures 4.4 and 4.5 show upper estimates of the change to wintertime and summertime maximum temperatures, respectively. With changes in winter temperatures of up to 6 °C and summer temperatures of up to 12 °C, the seasons of the future will be very different from those we remember or experience currently.

With the newer UKCP18 data, we are able to explore the changes to an event with a specific return period rather than just the warmest day or coldest day. Figure 4.6 shows the possible increase in summertime peak temperatures for a 1 in 100-year heat event. The data is shown for Islington in Central London (UK). While the upper end of the plume plot shows exceedingly high temperatures, we should note that this is a 1 in 100-year event and that in August 2003 temperatures of ~39 °C were recorded in London and more recently 40 °C in July 2022. Both of these heat events were greater than a 1 in 100-year return period and sit above the expected range shown in figure 4.6. It is easy to dismiss plots showing such high temperatures as fanciful; however, the risk is real, albeit with a large range of uncertainty, as indicated by the width of the plume plot by the end of the century.

While figure 4.5 indicates that the greatest change in summertime peak temperatures will be experienced along the South coast and in the Southwest of the UK, the greatest temperatures will still be experienced in the Southeast. Figure 4.7 shows possible peak summertime temperatures across the country for a 1 in 100-year

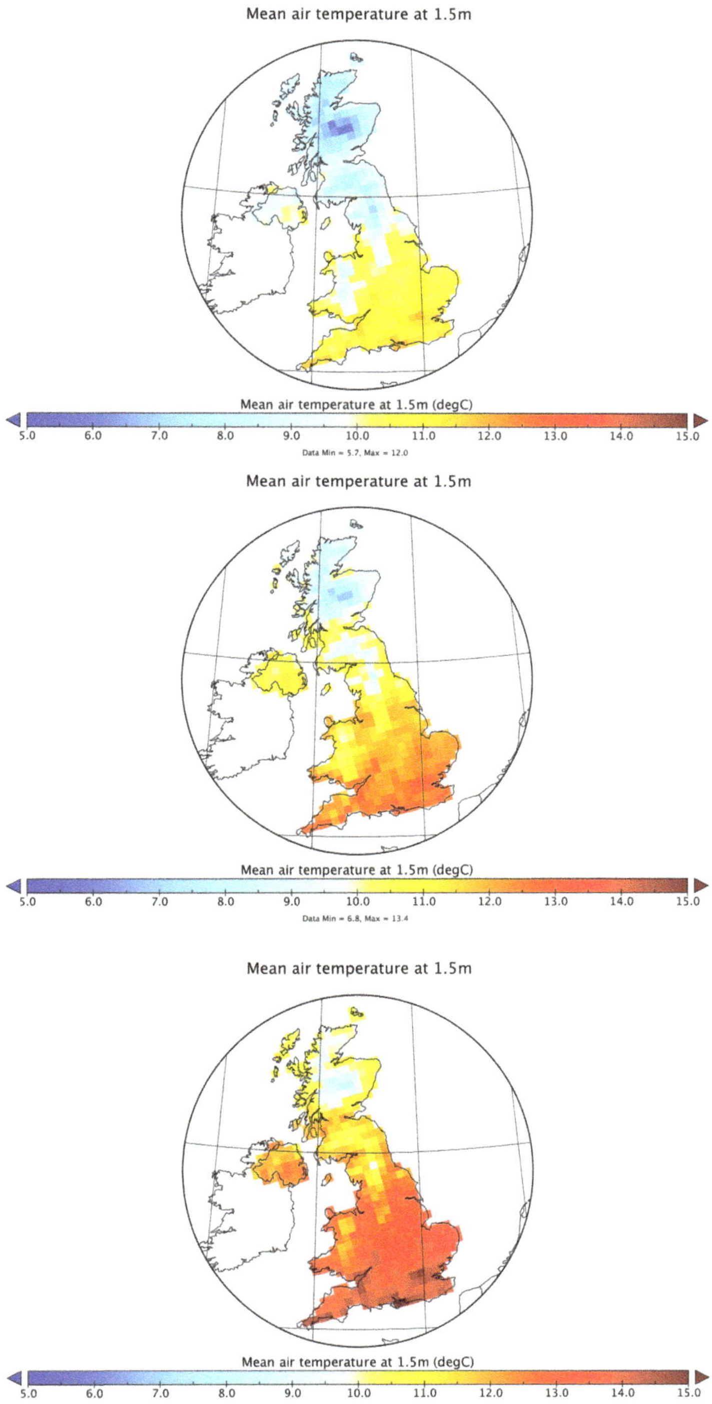

Figure 4.2. Mean annual air temperature for the UK for the 2020s (top panel), 2050s (middle panel) and 2080s (bottompanel) (data: UKCP09 A1FI 50th percentile under Open Government Licence v3.0).

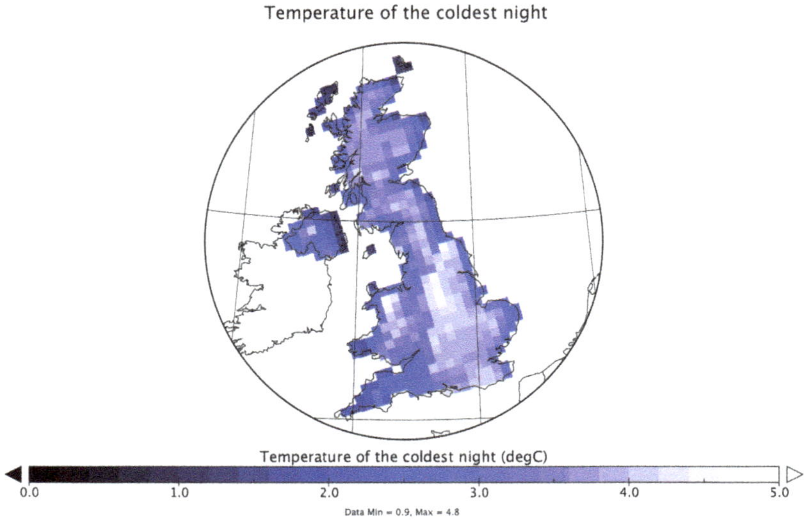

Figure 4.3. Change in temperature of the coldest night by the end of the century (data: UKCP09 A1FI 50th percentile under Open Government Licence v3.0).

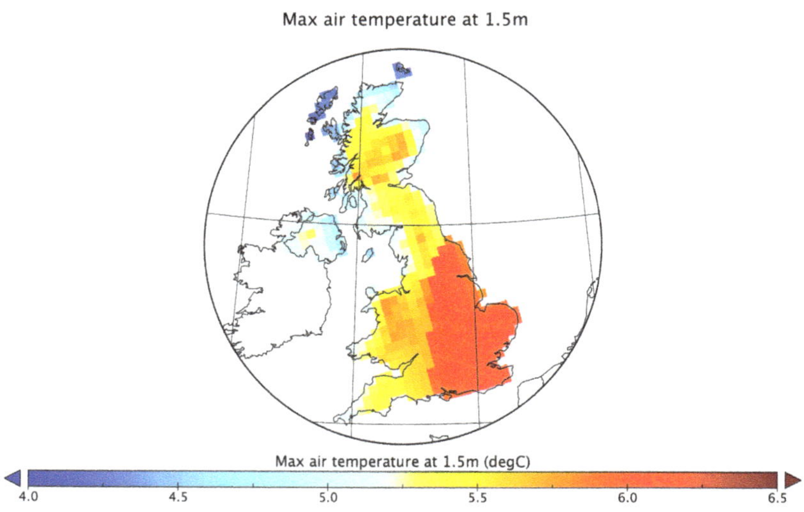

Figure 4.4. Change in average winter maximum temperatures by the end of the century (data: UKCP09 A1FI 90th percentile under Open Government Licence v3.0).

event by the end of this century. Such high temperatures will pose a significant risk to human life.

4.2 Human physiology and thermal comfort

Since the primary purpose of buildings is to provide shelter from the elements and thermal comfort for the building occupants, we need to consider how this may be

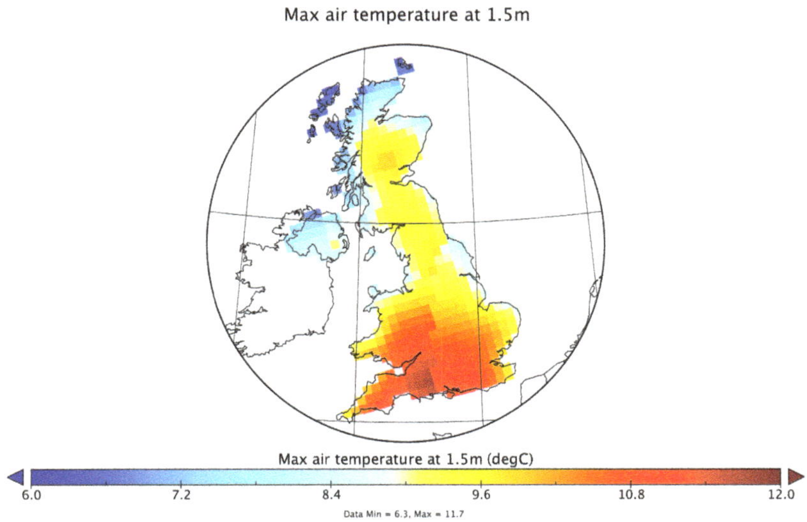

Figure 4.5. Change in average summer maximum temperatures by the end of the century (data: UKCP09 A1FI 90th percentile under Open Government Licence v3.0).

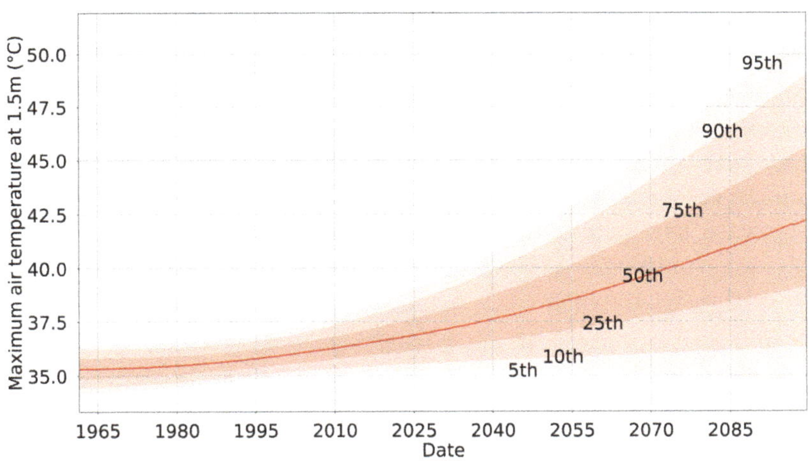

Figure 4.6. Plume plot of possible changes in summer maximum air temperatures over time for a 1 in 100-year event with different percentiles shown for Central London (data: UKCP18 RCP8.5 scenario against a baseline of 1981–2000 under Open Government Licence v3.0).

impacted by climate change. We will discuss changes to the design and construction of buildings to resist weather in chapter 8, but with regards to how to maintain the thermal comfort of the occupants in a changing climate we first need to consider what drives our perception of comfort.

The human body is subject to the laws of thermodynamics, we generate heat from energy in food and we need to lose this heat to our surroundings to maintain an

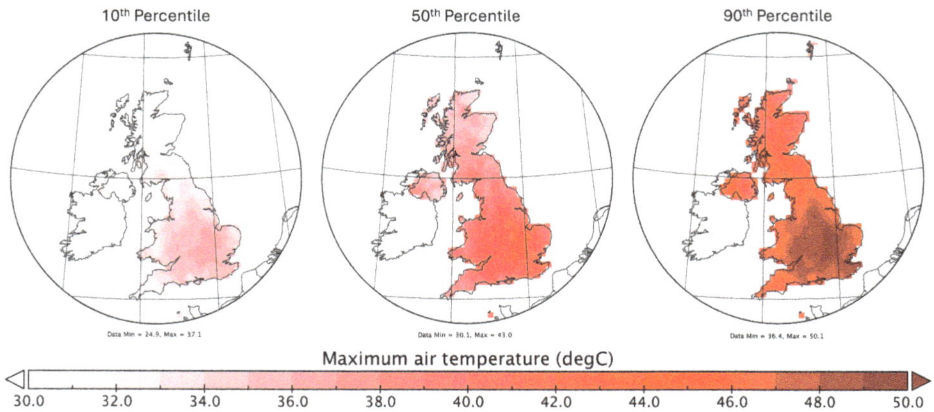

Figure 4.7. Plot of summertime maximum air temperatures by 2099 for a 1 in 100-year event for different probability levels (data: UKCP18 RCP8.5 scenario against a baseline of 1981–2000 under Open Government Licence v3.0).

internal temperature of 37 °C. If we did not lose this heat to the surrounding environment, our core body temperature would increase by about 1 °C h^{-1}, leading to death soon after. The mechanisms our body uses to control our body temperature are dependent upon the surrounding environmental conditions and the clothes being worn, clothing acts as insulation around our bodies limiting heat loss. Our body has several physiological mechanisms to control our body temperature, such as dilating and contracting blood vessels. In this chapter, we shall focus primarily on heat loss rather than heat conservation.

The human body is not in a steady state, with the body constantly trying to find a balance between heat gain and heat loss depending upon the level of activity and environmental conditions. As such, we can introduce a conceptual heat balance equation for this dynamic system:

$$(M - W) = E + R + C + K + S$$

This heat balance equation considers the metabolic rate (M) of the person in question, which produces energy to do mechanical work (W) (e.g. climbing stairs), the remainder of which ($M-W$) is released as heat. This is balanced by heat lost or gained via evaporation/condensation (E), radiation (R), convection (C) or conduction (K). If the heat gains and losses are in equilibrium (figure 4.8), then a core body temperature can be maintained; however, if gains are greater than losses then heat is stored in the body (S) and the core body temperature increases (or falls if losses > gains).

The conceptual heat balance described above is a useful starting point but not much use if we actually want to consider the environmental impact on human physiology. In order to make this useful, we need to consider the specific values for heat production and heat exchange for a human body. Estimates of metabolic heat production are typically provided in Met for different activities. 1 Met = 50 kcal $m^{-2}h^{-1}$ = 58.15 W m^{-2} (surface area of the body) and is representative of the heat

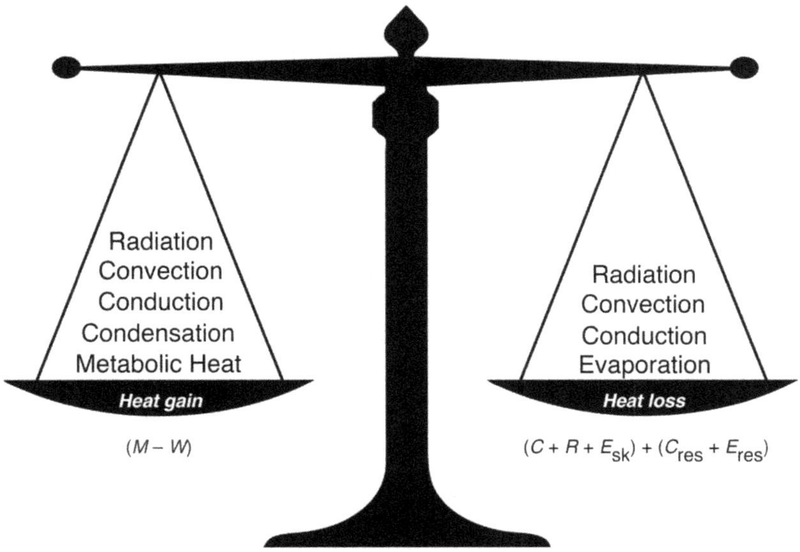

Figure 4.8. The conceptual heat balance—if both sides are equal, then a core body temperature can be maintained.

output of a seated person, further estimates of metabolic rate for different activities are presented in the table below:

Activity	Estimate of metabolic rate ($W\,m^{-2}$)
Lying down/sleeping	45
Sitting	58
Standing	65
Writing/typing	85
Walking on the flat at 2 $km\,h^{-1}$	110
Painting	130
Laying bricks	150
Carrying a 10 kg load on the flat at 4 $km\,h^{-1}$	185
Walking on the flat at 5 $km\,h^{-1}$	200
Loading a wheelbarrow with rocks	275
Cycling on the flat at 20 $km\,h^{-1}$	360
Going up stairs	440
Running on the flat at 15 $km\,h^{-1}$	550

In order to relate this to a specific individual, we can calculate the surface area of a person from their height and weight using the DuBois surface area (A_D):

$$A_D = 0.202\,W^{0.425}H^{0.725}$$

where W is the weight of the person (kg) and H is their height (m). Typically, values of 1.8 m^2 and 70 kg are used for men and 1.6 m^2 and 55 kg for women. The surface area of the person is necessary in order to calculate the available area for heat exchange with the surroundings, and in turn how comfortable or uncomfortable we are doing certain activities in different conditions.

The Danish scientist Povl Fanger carried out extensive work on the human perception of thermal comfort and how this is related to metabolism and environmental conditions. Detailed in his 1970 book *Thermal Comfort: Analysis and Applications in Environmental Engineering*, his work is still the primary metric for assessing comfort nearly 50-years later. He used a modified version of the heat balance equation:

$$M - W - E_{\text{dif}} - E_{\text{sw}} - E_{\text{res}} - C_{\text{res}} = R + C$$

where:

E_{dif} = heat loss by water vapour diffusion through the skin
E_{sw} = sweat rate
E_{res} = latent respiration heat loss
C_{res} = sensible (dry) respiration heat loss

Which can further be rearranged into the form shown in figure 4.8:

$$M - W = (C + R + E_{\text{sk}}) + (C_{\text{res}} + E_{\text{res}})$$

where E_{sk} is the total evaporative heat loss through the skin (both sweating and vapour diffusion). In this equation, we have two distinct heat loss paths—via the skin and via our breath. This is a useful way to view this heat balance because it makes it clear how we can impact different heat loss paths. Wearing more clothes (for example) insulates our body, this will reduce our outer surface temperature (the temperature of the clothing in this case not the skin), reducing heat loss by radiation and convection, but also reduces the ability of our skin to lose heat by diffusion or sweating. Increased clothing, however, will have no effect on heat loss via our breath. Fanger went on to quantify how each of the variables in our heat balance equation is affected by different environmental conditions, such as air temperature, the temperature of surrounding surfaces (radiant temperature), humidity, air velocity, amount of clothing, posture and level of activity.

Thermal comfort is often defined as ‘that condition of mind which expresses satisfaction with the thermal environment’. Since no-one has ever gone into a room and thought this room is (thermally) comfortable, we should consider thermal comfort as a lack of noticeable discomfort. A person’s perception of what makes them uncomfortable is highly complex, depending on individual metabolism, clothing, age, weight and ethic origins. But is also highly subjective, the reasons why a person reports thermal comfort (or discomfort) or related reports of warmth, freshness, stimulation or pleasure are not well understood. There is no single solution to determine if someone will be comfortable or not, and it is widely accepted that you cannot please everyone all the time. Instead the emphasis is on what environmental conditions will produce thermal comfort for the majority, based upon six basic parameters: air temperature (t_a), radiant temperature (t_r), air velocity (v), humidity (rh), clothing level (clo) and metabolic heat ($M-W$).

Fanger developed a system for quantifying how satisfied or dissatisfied a person is with their local environment, based upon the rated thermal sensations of a large group of test subjects. The subsequent metric predicted mean vote (or PMV) is still widely used and features in many building codes worldwide to identify suitable internal environmental conditions in buildings. PMV is ranked on a seven-point scale from −3 to +3, as shown below:

Hot	+3
Warm	+2
Slight warm	+1
Neutral	0
Slightly cool	−1
Cool	−2
Cold	−3

The calculation of PMV considers all six basic parameters mentioned above, along with air pressure in order to calculate sweat evaporation from the skin, convective and radiative heat loss. The methodology is complex and requires not only the insulative properties of the clothing worn but also the percentage coverage of the skin. This allows the area for sweat evaporation to be calculated along with the temperature of the skin and the clothing, respectively. Calculation of PMV is now programmed into many building energy/thermal analysis software packages as a metric to assess building performance. Where this calculation fails, however, is that it cannot estimate discomfort from temperature gradients (e.g. hot head and cold feet), which may arise from temperature stratification in a room. Despite this, PMV forms the basis for many other calculations (e.g. productivity covered later) and other metrics, such as the predicted percentage dissatisfied (PPD):

$$\text{PPD} = 100 - 95\exp\left(-0.03353\quad \text{PMV}^4 - 0.2179\text{PMV}^2\right)$$

This related estimate of thermal discomfort is interesting in that it ranges from 5% to 100%, indicating that one person in 20 will always be dissatisfied with the local environment.

As you might expect then, the most powerful mechanisms we have to control our body temperature are behavioural. We can add or remove clothing, change our posture to increase or decrease effective surface area for heat exchange, open windows or fan ourselves, move to shade and so on. Our bodies also have a number of physiological mechanisms to help maintain our core temperature. Both our behavioural and physiological responses continually interact at a subconscious level to respond to changing environmental conditions. These physiological mechanisms include, amongst others, vasodilation and vasoconstriction of skin arterioles, to increase/decrease blood flow close to the skin to increase/decrease radiative and convective heat loss. Sweating increases evaporative heat loss at the skin's surface, this coupled with vasodilation can assist in cooling the blood, and hence the body core. In cold conditions, we shiver to heat ourselves. Shivering effectively increases our metabolic rate from a rest rate of 65 $W m^{-2}$ to around 200 $W m^{-2}$ or more. All these different responses work to thermally regulate our bodies, primarily to safeguard our health, but also to provide comfort. This thermoregulation, however, has an impact on our mental focus and levels of productivity, due to fidgeting, distraction, complaining, time off work etc.

Currently, there are no comprehensive policies for how to adapt existing homes, offices, schools, hospitals or care homes to higher temperatures. Both average and maximum temperatures are expected to increase across the UK over the next few decades. Previous extreme events, such as the European heatwave of August 2003, are expected to be representative of a typical summer by the 2040s. Such events have been shown to have a significant impact on the productivity of workers. The heatwave of 2003 is estimated to have cost the UK between £400–500 million in reduced manufacturing output over the two-week period.

There have been several studies of how environmental variables such as temperature affect the productivity of workers. It is generally accepted that there is an economic benefit for workplaces to maximise worker satisfaction. The question is what factors increase worker satisfaction in the work place? There is evidence that giving people personal control over temperature and ventilation improves perception of comfort and increases satisfaction. In hot environments, blood vessels in the skin will dilate to increase radiative and convective heat loss, and then increase sweat rate to increase latent heat loss. If these are insufficient, the core body temperature will rise. Increasing core body temperature raises heart rate and causes fatigue because the body must work harder to try and maintain its internal temperature. The resulting heat stress and discomfort will lead to behavioural changes and effects on cognitive performance, such as mental performance, information processing, memory and so on.

Many worker tasks require both physical and cognitive functions; for example, typing requires speed and accuracy while processing information. Being in thermal discomfort affects different aspects of tasks in different ways, as such it is necessary to consider a metric of thermal comfort such as PMV that accounts for all these factors and the level of activity being undertaken. Many office workers use computers for a significant part of the day. Computing is a special task requiring close attention to details and visual cues. The requirements of computers for the user to follow specific logic based procedures results in a high optimal level of arousal.

Moderate heat stress of a few °C above optimum can have an adverse effect on the performance of tasks due to relaxation and a reduction in arousal. This is considered to be an autonomic response to regulate body temperature at the limit of vasodilatory control, prior to the onset of sweating. It has been noted that a conscious effort can override this effect and maintain arousal levels. Where a person enters a hot room and is exposed to high temperatures suddenly, this can act as the cue to exert this conscious effort. However, in most cases temperatures will rise slowly over the course of the day, circumventing the opportunity to exert conscious effort, and resulting in lower arousal and concentration. Worker performance depends upon the level of arousal compared to that required for optimum performance. A task that is considered boring will be de-arousing and a subject will perform better if their arousal level can be raised by some means. Stimulation caused by a heat, cold or draughts may increase arousal but can also lead to distraction and time off task if arousal is increased too far.

If we are to consider how adverse thermal environments will affect worker performance, we need a relationship between a comfort metric such as PMV and worker performance. Roelofsen [1] compiled a direct relation between loss of performance and thermal comfort based upon several studies of worker performance and models of how humans respond to thermal load:

$$P = b_0 + b_1PMV + b_2PMV^2 + b_3PMV^3 + b_4PMV^4 + b_5PMV^5 + b_6PMV^6$$

where P is the loss of performance as a percentage and the values $b_{0\text{–}6}$ are regression coefficients, depending on whether you are on the hot or cold side of neutral. Roelofsen's relationship between productivity and PMV can be seen in figure 4.9. The red data point indicates thermal neutrality (PMV = 0), we can see that there are maxima in relative productivity either side of neutrality, with the greatest being on the cool side. This is to be expected, as discussed previously an increase in arousal due to slight discomfort will improve productivity.

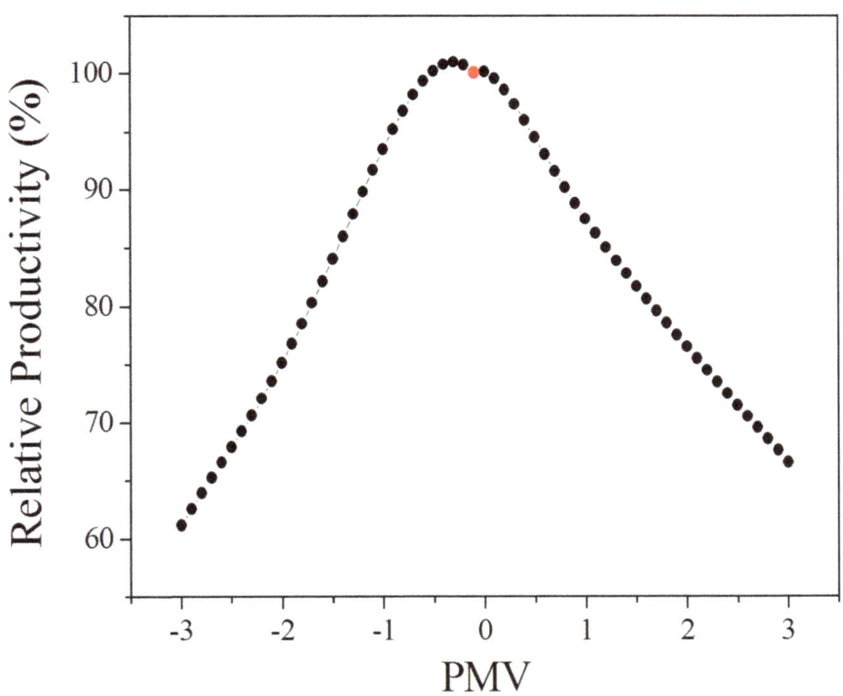

Figure 4.9. Plot of the relationship between productivity and thermal comfort.

In addition to reduced productivity, increased temperatures and the resultant heat stress from trying to maintain a core body temperature can have health implications. If heat needs to be dissipated, then the body's central blood volume decreases so that blood vessels near the skin can dilate (increasing volume). This reduces the stroke volume of the heart, which must then beat faster in order to maintain blood volume flow rate to the bodies organs. The effective circulatory blood volume will further decrease after the onset of sweating as water is lost from the body. If this water is not replaced, then the heart must work harder (as well as faster) to pump the more viscous blood around the body. If sweating persists for a prolonged period, the body can become dehydrated, which can lead to further physiological stress and illness.

The impacts of hot weather events such as heatwaves and warming from climate change resulting in overheating in public buildings such as hospitals, care homes, schools and prisons across the UK are not currently known. This is because high temperatures can have greater impact on the very young and elderly, whose bodies are unable to regulate temperatures as effectively. In addition, infirm and vulnerable groups are particularly susceptible, and it is currently not well understood how different conditions, illnesses or even medications affect the ability of our bodies to control temperatures.

4.3 Overheating

Even for healthy, able people who can take behavioural steps to cool themselves and whose bodies are able to deploy the full range of physiological responses to high temperatures, there is still danger from prolonged exposure to high temperatures. It is currently unclear how quickly humans will adapt to a changing climate. If we are unable to adapt quickly enough, it is estimated that the annual UK heat-related mortality (currently around 2000 deaths per year) will increase by two-thirds in the 2020s, increasing to ~250% by the 2050s and by more than 500% by the end of the century. These estimates are based not only on a warming climate (a medium emissions scenario (A1B) was used) but also a growing, aging population. The percentage of people in the UK aged over 75 is expected to increase from 8% in 2015 to 18% by 2085. More elderly people are more likely to be adversely affected by high temperatures, due to factors such as reduced sweating efficiency, but at temperatures above 35 °C, all age groups are at risk. It is possible that humans will undergo some form of autonomous physiological acclimatisation to gradual increases in mean temperature. For instance, it is possible for someone to emigrate from one climate to another and over time become acclimatised to that climate. However, evidence suggests people are unable to adapt to sudden changes in temperature over a short period of time, particularly if overall temperature variability also increases, as is expected as a result of climate change.

Numerous definitions of what constitutes overheating exist, in the UK an example is TM52 and TM59 from the Chartered Institution of Buildings Services Engineers (CIBSE), which specify various metrics of temperature exceedance. These can take several forms such as: (i) a number of hours above a threshold temperature, or (ii) a daily weighted exceedance, which considers the severity of the overheating within any one day, or (iii) an upper limit which must not be breached over the course of a year. An alternative is the American Society of Heating, Refrigerating and Air-Conditioning Engineers (ASHRAE) standard 55, which uses a range of acceptable comfort using the PMV metric. The comfort range typically specified is $-0.5 < \text{PMV} < +0.5$, a value of PMV outside this range would be considered uncomfortable. Temperature based metrics work well for assessing overheating for buildings in 'free-running' mode with no mechanical control (e.g. air conditioning). These buildings are typically highly dynamic and the internal environment responds quickly to changes in the external conditions. The comfort-based metrics which consider other variables in addition to temperature are better suited to building with mechanical control where aspects such as humidity and air speed are also controllable. A considerable limitation with all these metrics, however, is the assumption that the specified temperatures and comfort ranges are suitable. Although these metrics are derived from many years' worth of studies of temperatures in buildings, they are typically from studies that are conducted in cooler climates in the Northern hemisphere (e.g. Northern Europe and North America). One should therefore be wary of using the same metric in different climate zones because the local populous may well be adapted to a different range of temperatures, making a universal assessment of the risk of overheating difficult.

Worryingly, there is evidence that people lack a basic understanding of the risks to health from indoor high temperatures, and as a result are less likely to take measures to safeguard their and any dependents' wellbeing. The UK Committee on Climate Change warns that insulating homes to improve winter thermal efficiency needs to be undertaken carefully to avoid increasing the risk of overheating. This is largely due to the way that we currently design our homes and is an indication that the climate change mitigation and adaptation agendas are not always complimentary. This is something that will need to be addressed over the next few decades to safeguard the population.

We can estimate the risk to human life from overheating using the summertime two-day mean external temperature ($T_{2-\text{day}}^{\text{mean}}$), which has been shown to be strongly associated with excess summer deaths. The relative risk of death increases linearly above a threshold temperature, coincident with the 93rd percentile of $T_{2-\text{day}}^{\text{mean}}$. Thus, we can estimate the heat-related mortality (M) over the summer from the relative risk (RR) given by:

$$M = D_{\text{summer}}^{\text{all-cause}} \times (\text{RR} - 1)$$

with

$$D_{\text{summer}}^{\text{all-cause}} = \frac{d}{365} \times D_{\text{year}}^{\text{all-cause}}$$

and

$$\text{RR} = \left(\alpha \times T_{2-\text{day}}^{\text{mean}}\right) + \beta$$

where $D_{summer}^{all-cause}$ is the deaths over the summer from all causes, d is the number of days when the T_{2-day}^{mean} is above the threshold temperature and α is the heat-mortality gradient in % per degree above the mortality threshold temperature (a measure of how adapted people are to higher temperatures). External air temperatures are not always the best indicator of for assessing risk to human health, instead temperatures inside buildings are more appropriate. It follows then that the internal mortality threshold temperature would be the 93rd percentile of internal temperatures. This assessment allows a comparison between different types and constructions of buildings and their resilience to higher temperatures. Liu *et al* [2] mapped the overheating risk and related summertime mortality for different building types and constructions across a medium-large city in the UK (Sheffield, population ~553 000), for the 2020s (2010–39) and the 2050s (2040–69). The external mortality threshold temperature for Sheffield has been shown to be 22.2 °C with α equal to 1.7%. β can be calculated when RR equals one and T_{2-day}^{mean} is equal to the mortality threshold temperature.

It is currently unclear at what rate people will adapt to the higher temperatures expected in the future. If people do not adapt quickly, the mortality threshold temperature for the 2050s will be similar to the 2020s; however, if people do adapt, then it follows that the mortality threshold for the 2050s will be the 93rd percentile of 2050s temperatures, and any increase in heat-related deaths will be due to changes in the shape of the temperature probability distribution function. Maximum temperatures are expected to increase more than minimum and mean temperatures, implying that the distribution will become wider and more skewed.

As stated above, the heat-related mortality M is determined from the linear relationship between relative risk (R) and the internal two-day mean temperature (T_{2-day}^{mean}) above a citywide mortality threshold temperature. The citywide mortality thresholds for Sheffield were calculated as 24.4 °C and 25.9 °C for the 2020s and 25.4 °C and 27.2 °C for the 2050s for the 50th and 90th percentiles respectively. Figure 4.10 shows the results of Liu *et al*'s [2] simulations showing mortality maps of Sheffield at a ward level. Variation in mortality rate between wards is due to differing altitudes and different building constructions.

Average projections of citywide M for the 2020s, at the 50th and 90th percentiles are 7 and 12 per million per year, respectively, increasing to 21 and 39 per million per year in the 2050s in the absence of any adaptation (using the 2020s mortality threshold temperature). Thereby, indicating that the heat-related mortality rate would roughly triple in a 30-year period if people were unable to adapt to a warming climate. This is in line with estimates from the committee on climate change. If people do adapt quickly enough to increasing temperatures (using a 93rd percentile threshold temperature for the 2050s), the adverse impact of the warming climate on M is largely offset by human physiological adaptation. The mortality rates shown in figure 4.10 are comparable with those recorded in other UK cities during hot periods of weather. What is clear from the work of Lui *et al* is that the type and construction of the dwelling is important. It was found that terraced and semidetached dwellings showed the greatest amount of overheating particularly at night. This was due to the combination of mainly solid brick walls with poor insulation (which can keep heat

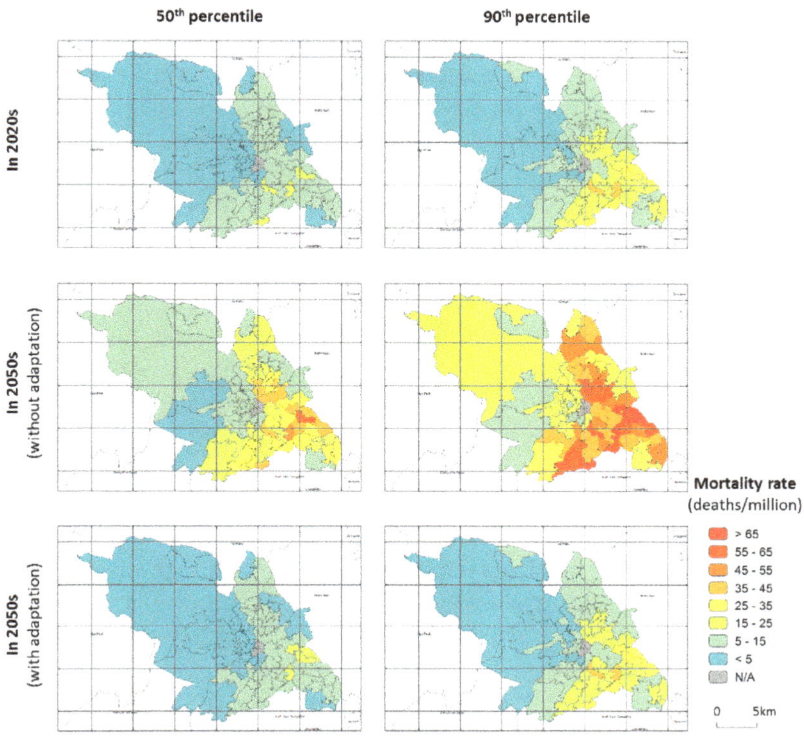

Figure 4.10. Map of heat-related mortality at a ward level for Sheffield for the 2020s and 2050s for different levels of climate change. Reprinted from [2] Copyright (2017), with permission from Elsevier.

out as well as in), with the high thermal capacity material (bricks) exposed to direct solar radiation. This allows the bricks to reach a high temperature and the heat to enter the building relatively easily. Finally, both terraced and semidetached homes have a reduced surface area over which to lose heat by radiation and convection due to the party-walls between dwellings, detached houses showed comparatively lower overheating risk despite similar building constructions. Interestingly, bungalows showed high daytime temperatures but the lowest level of nocturnal overheating in the study. This was attributed to the large roof area compared to the volume of the building. Since all rooms are on a single floor in a bungalow, the area of roof exposed to solar radiation is much larger than for a two storey (or greater) dwelling. Therefore, there is a large area to absorb heat from the Sun, which is several hundred W m^{-2} depending on time of day, azimuth and inclination of the roof. However, this also means that there is a large surface area over which to lose heat at night, meaning that the bungalows cooled faster and showed less nocturnal overheating. It seems clear then that there is the potential to adapt the way we design our buildings and urban areas to reduce heat storage and overheating, thereby reducing the risks to human health and increasing task productivity.

4.4 Overcooling

Unlike overheating—which is the occurrence of internal temperatures above the thermal comfort range, the cause of which is usually as a result of high external temperature, high solar gains through windows or uninsulated building elements or even as a result of excessive internal heat gains (people, lighting and heat emitting equipment), all of which are non-intentional—, overcooling is the deliberate reduction of temperatures below the comfort range.

As we have discussed previously, the global buildings sector is responsible for about 30% of global energy consumption and its associated carbon emissions. Space cooling is a significant energy use within buildings, responsible for roughly 20% of the global final energy consumption and about 8% of the carbon emissions in the buildings sector. As one might expect, cooling is more prevalent in warmer climates but it is also common in cooler climates, particularly in building with large internal gains (e.g. spaces with a large number of people or computing equipment such as offices). Within warmer climates, space cooling is the largest but also the fastest growing energy end-use. The vast majority of growth in global built floor area will be within warmer climates and the majority of this growth will occur over the next few decades. As such, global space-cooling energy consumption is projected to triple from $\sim$2000 TWh currently to >6000 TWh by 2050, the greatest part of this growth will be in domestic buildings.

It may seem remarkable that a significant proportion of the projected growth in space-cooling demand is being driven not by the growth in the number of air-conditioning units, and hence greater internal thermal comfort for the warmer (and often less developed) regions of the world, but instead to reduce internal temperatures below the comfort threshold. This unnecessary expenditure of energy to produce uncomfortably cool conditions is termed overcooling. Unmanaged overcooling will not only lead to the wasted expenditure of energy within a rapidly developing world but as the built floor area grows it will also place increased burden upon energy supplies and hamper efforts to reduce energy sector carbon emissions due to increased peak loading during hot weather (see chapter 3). There is also the issue that thermal discomfort reduces worker productivity and that prolonged periods away from thermal neutrality can worker health and increase worker absence.

This effect has gained some notoriety amongst buildings, such as offices and shopping centres, with the term 'air-conditioningitis' being used to describe the negative health effects of the large disparity between internal and external temperatures caused by low set-point temperatures in these buildings. This practice of overcooling drives occupants to adjust their practices to remain thermally comfortable, suich as by wearing extra clothing inside or even energetically wasteful measures such as opening windows to dump the cool air and let warmer air inside.

So, what causes this wasteful overuse of cooling? Two main causes have been identified, being either due to cultural factors or buildings regulations based. There is evidence that in some hot climates that cooling is seen as a statement of wealth and hence there is a desire to cool your house/office as much as possible to demonstrate this despite the thermal comfort implications. The second reason can be due to the fact that human thermal comfort criteria were derived and tested using healthy, middle-aged people from Northern Europe and North America. These areas typically have a temperate climate and the expression of what constitutes thermal neutrality will be based upon people adapted to living in this temperate climate. While there are set upper and lower limits for human thermoregulation, the method by which the human body achieves thermoregulation can vary between people adapted to different climates. For instance, in warm humid climates sweating is less useful as a heat loss mechanism, so vasodilation tends to be greater, while in hot dry climates the opposite would be true. This has implications for set-point temperatures. People living in hot climates will have become naturally adapted to maintaining thermoregulation in these climates, and as such may be tolerant of higher temperatures. This becomes an issue when we consider that many countries still do not have their own building codes and instead adopt existing codes such as those from ASHRAE which stipulate a range of suitable temperatures or ranges of acceptable PMV. These temperature and comfort ranges will be based upon and designed for those living in temperate climates, potentially leading to overcooling. Since no universally accepted definition of thermal comfort exists, some countries such as India have chosen to define their own standard rather than opt for using the ASHRAE Standard 55, ISO 7733 or the European Standard EN16798.

Recent research [3] has examined the prevalence of overcooling in offices in across a number of countries with warm/hot climates and has examined the energy implications of overcooling. This research used assessments of thermal sensation and thermal preference votes (TSV and TPVs) recorded in the ASHRAE Thermal Comfort Database, which contains data from 90 000 occupants of office buildings located across 27 countries. From the TSV and TPV data, we can assess how hot or cold people feel but also how much warmer or cooler they would like to be. This data coupled with environmental monitoring and a calculation of the associated PMV of the internal conditions allowed an assessment of the level of overcooling present across the ASHRAE database. The use of both TSV and TPV is considered the safest way of determining occupants' attitudes towards thermal comfort and overcooling because it uses both the subjective assessment of comfort and also the preference, thereby removing the possibility of bias, ambiguity or misinterpretation of the criteria. When these results were compared to the calculated PMV for warm climates, there was a clear discrepancy between PMV and the occupant responses which favoured a warmer internal environment again highlighting the fact that there is no universal thermal comfort index.

The work of Alnuaimi *et al* [3] observed the average temperature of offices in warm climates was be 24 °C, which is equivalent to a PMV of +0.08, indicating the need for more cooling, while the comfort temperature of the occupants derived from the PSV and TPV results was calculated to be 25.5 °C on average. This analysis not only highlights the clear difference in observed temperatures to comfort temperatures but also the range of variability between buildings and climates. The calculated comfort temperature varied from 0.4 °C to 3 °C above the observed temperature (with an average of 1.5 °C). Interestingly, a significant disparity was observed between the cold discomfort reported by males and females, with females experiencing approximately three times higher cold discomfort compared to males. This could be due to a number of reasons, including lower metabolic rates and activity levels or due to differences in the insulation properties of clothing. This reinforces the need to consider comfort and discomfort across genders, cultures and climates, rather than adopt a one-size-fits-all approach.

The thermal discomfort caused by overcooling will have a knock-on effect for occupant health and wellbeing. Common health problems include cold-like symptoms, increased fatigue and headaches. These issues can worsen with prolonged exposure or underlying illness or advanced age. Cold discomfort is typically associated with a loss of productivity (see figure 4.9), disruption to fine motor skills due to vasoconstriction and increased viscosity of synovial fluid in the digits. Prolonged discomfort and its associated impacts can lead to increased absenteeism and time off work. This amounts to a large financial cost not only for the increase energy expenditure but also loss of income and wasted staff costs which can account for as much as 90% of total costs.

The impact of overcooling is not only about occupant comfort it is also about the unnecessary expenditure of energy, and hence carbon emissions. Globally, fossil fuels still account for approximately 65% of total final energy consumption, with an average carbon emission factor of 0.505 $kgCO_2/kWh$ in 2016 according to the IEA. This value is typically higher for warm climates, especially those with large fossil fuel reserves. Although there are trends towards lower carbon intensive energy generation, a significant increase in the demand for cooling is expected as a result of climate change but also due to increased urbanisation, economic growth and development in the Global South.

Given the challenges identified in chapter 2 regarding reducing energy consumption in the buildings sector and adopting a completely renewable energy system, the wasteful consumption of energy to overcool buildings is of great importance if allowed to grow unchecked, in line with the expected growth in cooling. As mentioned earlier, space-cooling currently accounts for approximately 2000 TWh of global energy usage, or equivalently about 8% of total global carbon emissions. As a simple approximation, if we assume that 17% of the global cooling energy consumption is wasted on overcooling as indicated by Alnuaimi *et al*'s [3] research, then we have about 1.3% of global carbon emissions is currently being emitted unnecessarily. Looking forward, this 1.3% currently, which may seem inconsequential, will grow. By 2050 the 2000 TWh of cooling expenditure will grow to $\sim$6000 TWh, and most of this growth will be in warm climates located in the Global South. These countries are unlikely to have decarbonised their energy grids to the same extent as the industrialised parts of the world. As such, the proportion of carbon emissions due to overcooling can be expected to grow considerably. Since we have assumed that overcooling is the purposeful expenditure of energy to cooling buildings below comfort levels, it has the potential to be mitigated simply by turning up the cooling set-point temperature. This, however, will require an understanding by building practitioners and occupants and needs to be captured in local buildings guidance and codes because the one-size-fits-all approach currently adopted is unsuitable.

4.5 Building physics and possible adaptations

Overheating in buildings is clearly a pressing issue. However, there is evidence that as we attempt to make our buildings more energy efficient by increasing levels of insulation and standards of air tightness, we are also increasing the risk of overheating. The National House Building Council has collated several sources of evidence regarding the construction of new homes, the refurbishment of old ones and the risk of overheating. Historically, our homes have been designed to keep heat in to provide thermal comfort in winter, with only small openings for ventilation and no solar shading. As we have made our buildings better at keeping heat in and simultaneously increased the amount of heat generated in our homes, we have not fundamentally changed the way we design or use our buildings so overheating seem inevitable. Our homes are smaller than ever before and we have more heat generating equipment within them, such as more TVs, computers, tablets, phones, set top boxes, home entertainment systems, etc. All of these generate heat, and yet we have not increased provision to remove this heat from our buildings in the summer months. We also have not added solar shading to reduce solar radiation entering through windows in the summer months. Double glazed windows are now the norm in the UK, each pane of glass only reduces the amount of solar radiation (visible and near-infrared) entering the building by a few percent being transparent, but reduces the heat (thermal infrared) that can pass back out again by about half (glass is opaque in the thermal infrared). Without shading, the presence of windows placed on façades to provide light airy rooms that we favour in the UK can easily lead to high temperatures and heat that cannot escape from our buildings due to high levels of insulation and poor ventilation provision.

The materials used to construct our buildings and urban areas are extremely important because they are what controls the exchange of heat between the internal environment and the outside world, via radiation, convection and conduction. The emissivity ε (a measure of how absorbing or reflective an object is) of the materials used is especially important in sunny locations. Radiation striking an opaque surface can be reflected or absorbed (usually a mixture of the two) depending upon its ε. The colour of a surface gives a good indication of the ε in the visible spectrum. The lighter the colour, the more reflective with a lower ε, while darker surfaces are more absorbing with an ε closer to 1. Hence, darker surfaces absorb more solar radiation and will attain higher temperatures than lower ε surfaces. The ε of a surface is also a measure of how radiative a surface is at different wavelengths, with the most absorbing surfaces ($\varepsilon \approx 1$) also being the most radiative. The colour of a material, however, tells us nothing about the ε of the material in non-visible wavelengths of light, such as the near-infrared or thermal infrared. Paint, for example, can be tailored in the visible spectrum to different colours, and hence different emissivities; however, due to the carbon in the paint, all colours are highly absorbing/radiative in the thermal infrared. Hence, both white and black painted surfaces emit heat equally

Figure 4.11. The many-coloured buildings of Elba (Italy).

at night even if their absorption of solar radiation differs during the day. As such, light-coloured paints have been used for centuries in warmer locations to help keep buildings cool. The Italian island of Elba (figure 4.11) is a prime example of this phenomenon—the buildings covering the island come in a wide variety of colours, but all are light in hue with a low ε in the solar spectrum.

The buildings of Elba and of many other Mediterranean countries employ light-coloured painted façades to reflect the solar radiation away from the building. This, however, is not enough to maintain thermal comfort. Buildings in warmer climates are typically made of heavyweight materials, which can remove heat from the air during the day to be lost at night when air temperatures are lower.

Heat storage within a material depends upon its mass m (kg) and its specific heat capacity c_p (kJ kg^{-1} K^{-1}). The ability of a material to store heat within a building is often termed thermal mass—the greater the specific heat capacity or the greater the mass, the more heat Q can be stored:

$$Q = c_p m \Delta T$$

where ΔT in this case is the difference in temperature between the material and the surrounding air. The specific heat capacity c_p for many typical building materials are relatively similar, even between those used as insulation and those considered lossy. For a material to be considered as good, thermal mass it needs to be relatively conductive to heat to allow it to maintain a uniform temperature, else the surface would simply warm up to meet the air temperature and heat transfer would cease. This also imposes the limitation that the mass should be relatively thin, else heat would not have time to travel through the material to access all available mass. This implies that dense materials are the best thermal mass, so that there is a substantial mass that can be accessed by heat travelling by conduction from the surface. As such, thermal mass can be quantified by the volumetric heat capacity of a material (kJ m^{-3} K^{-1}), a measure of how much heat can be stored per unit volume. Materials such as stone and concrete are good examples of thermal mass.

The greater the thermal mass of a building, the greater the effect it has on moderating the external air temperatures. The high specific heat capacity and density of thermal mass means that a large amount of heat can be stored in the material, this implies that the surface temperatures will be lower than a low thermal mass material. A lower surface temperature reduces heat lost by the material through convection and radiation. Reducing the rate of heat loss means that the heat stored within thermal mass takes longer to be released, shifting the diurnal temperature cycle and offsetting peak temperatures. Figure 4.12 shows an example of internal air temperatures for both a thermally lightweight building (e.g. timber frame or aerated blockwork structure) and a thermally heavyweight structure (e.g. cast concrete, brick or blockwork) compared to the external air temperature. There is still a lag in time and a reduction in the peak temperature for the lightweight building due to the insulative properties of the structure but these are minimal compared to the effects of the heavyweight structure. With a thermally heavyweight building, internal temperatures can be significantly reduced by several °C and the extremal temperatures offset by several hours. This can be particularly useful (for example)

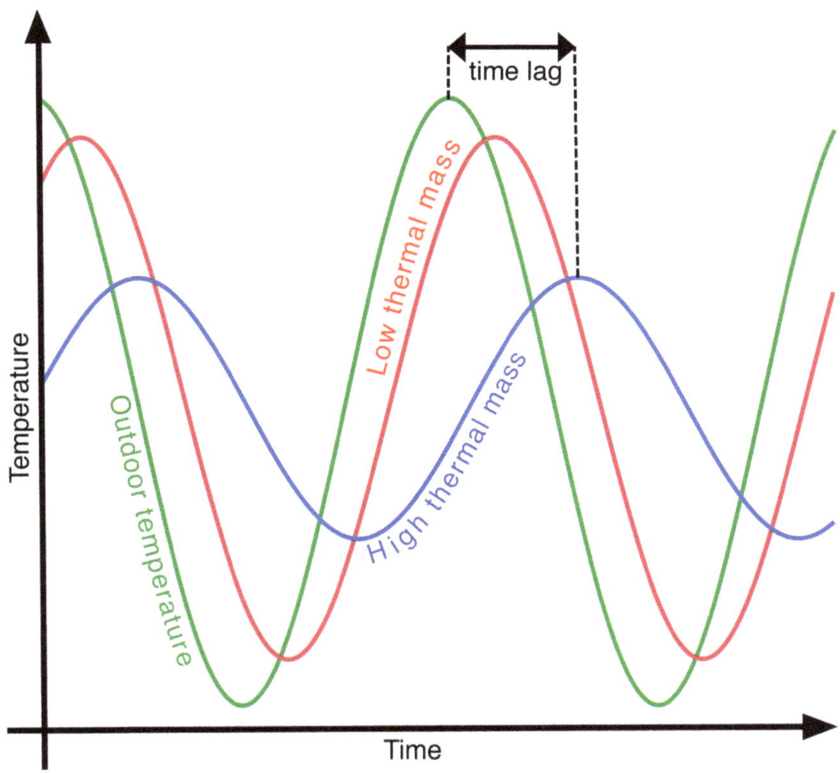

Figure 4.12. The relationship between time and temperature for different building constructions.

for an office building where significant heat is generated during the day from people and equipment, as well as any solar gains. A heavyweight structure will allow this heat to be stored in the structure to be released after the working day has ended, keeping the internal environment cool during working hours, reducing or even avoiding the need for air conditioning.

Part of the reason why new build homes are at risk of overheating, in addition to improved insulation and air tightness, is the choice of materials and the construction methods now used. There is now a prevalence in the UK to construct new buildings out of materials that are quick and easy to use, thereby saving time and money. Where once internal walls would have been brick, now plasterboard and timber stud work is common. While brick and block outer walls have been replaced with lightweight timber frame or aerated concrete block walls. Perhaps the biggest change is the move away from wet plaster finishes to the room interiors, with plasterboard being used instead. Unlike solid plaster, plasterboard is more insulative and is not in direct thermal contact with the wall construction behind due to a small air gap. This has the effect of masking any thermal mass that may be in the wall behind, making the building act as a thermally lightweight structure, even if heavyweight materials have predominantly been used. A similar effect can be seen in modern office buildings, which are often highly glazed, meaning the majority of the thermal mass is in the concrete floor slabs between storeys. However, on one side this is covered with carpet or other insulative floor finish, while the underside of the concrete floor slab is masked by a suspended ceiling, to hide services and improve room acoustics. Limiting thermal mass or masking it behind an insulative material can, however, be beneficial in some cases. The presence of a significant amount of thermal mass makes the building very slow to respond to changes in environmental conditions, this also means that thermally massive buildings are slow to warm up when the heating comes on or to cool down when a window is opened or when the air conditioning is turned on. Hence, there is a trade-off between peak temperatures and responsiveness to building controls, which needs to be considered when designing a building.

Buildings are not just stand alone entities, they exist in thermal balance with their surroundings. Hence, when considering how to improve the thermal performance of a building, we also need to consider the surrounding context. As we discussed in Chapter 1, solar radiation heats the Earth's surface. The resultant temperature depends upon the latitude, season, time of day, the inclination of the ground and the type of terrain (figure 4.13). Natural vegetative cover of terrain moderates extreme temperatures and stabilises conditions through evapotranspiration, increased albedo (reflectivity) and shading of the ground below. Plants and grassy terrain cover tend to reduce surface temperatures but cities and man-made surfaces tend to increase surface temperatures, this will be covered further in Chapters 5 & 6. An indicative heat balance for different surface types is shown in figure 4.13. A common mistake which can lead to unpleasant environmental conditions is to place paved surfaces, which store a large amount of heat due to their density and specific heat capacity, close to the windows of buildings. These paved areas not only have the drawbacks discussed earlier in Chapter 3 but also reflect a significant proportion of solar

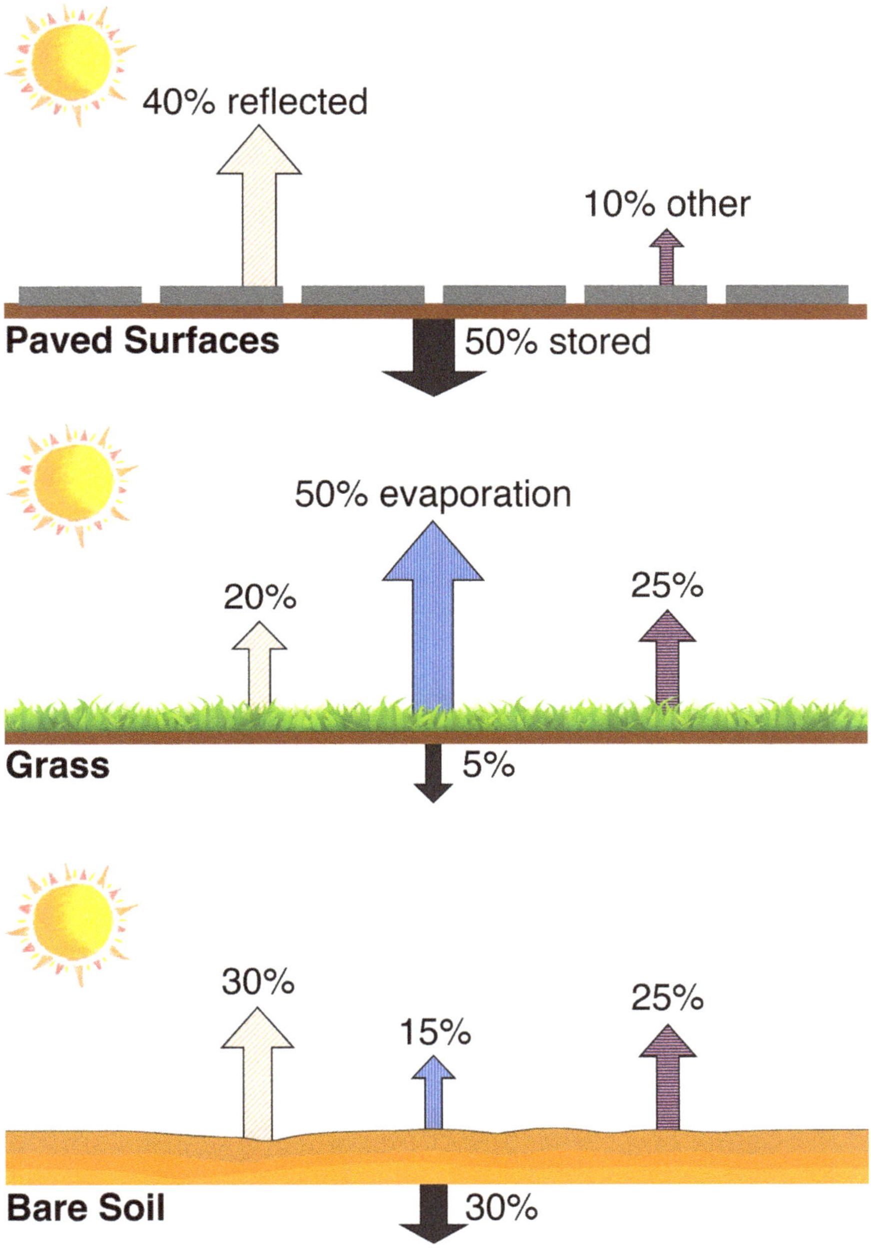

Figure 4.13. Heat balance for different surface types, showing proportion of reflected solar radiation, heat lost by evaporation or radiation/convection/conduction (shown as other) and heat stored in the ground.

radiation, which can then enter buildings via windows. Roughly half the solar radiative power is converted into heat, warming the paved surface. Due to the thermal mass of the paving a large amount of heat can be stored, meaning that the paving will still be hot and radiating a large amount of heat after air temperatures have fallen. This can have an adverse effect in summer evenings when windows are open and we are trying to purge our dwellings of heat because the paved surface will not only heat the local air but also radiate heat directly into our building preventing it from cooling effectively. By comparison, grass is a much cooler surface due to a large proportion of the solar power going into the evapotranspiration of water from the soil surface and through the stomata of the blades of grass. Even bare soil, due to its permeable nature, allows for solar power to be used for the evaporation of moisture from the soil's surface. This is true for other permeable surfaces as well, manifesting as a reduced surface temperature (figure 4.14), providing an addition benefit for the Sustainable (Urban) Drainage System (SuDS) systems discussed in chapter 3.

Our experience of heat in a building is affected as much by the heat radiated from the surfaces around us (the radiant temperature) as it is by the air temperature. So our perception of comfort can be improved in hot condition by having cool surfaces around us, this implies the use of high thermal mass materials to keep surface temperatures down. If you walk into a stone building, such as a church, on a hot day you immediately sense the change in the radiative environment. The human body is almost a perfect black body in the thermal infrared spectrum, and hence is very sensitive to changes in the amount of thermal radiation received. The colour of the stone is also important, the light colour reflects more of the solar spectrum than dark materials, reducing the amount of heat that the building absorbs. A dark wall exposed to the Sun can have a surface temperature 10 °C greater than a light-coloured wall in the same conditions. Figure 4.14 shows some indicative temperatures for different surfaces around an urban area exposed to sunlight. The choice of materials deployed in our buildings and urban areas can have a profound impact on our perception of thermal comfort by altering the mean radiant temperature.

The orientation of a building or street plays a major role in the thermal balance between a building and its surroundings, determining access to wind and receipt of solar radiation. In hot climates, for instance, protection from solar radiation is paramount, particularly during times of high temperatures (e.g. just after midday). When considering the optimum orientation of a building, there needs to be a balance between minimising solar receipt during the hottest months and solar receipt during the coolest months. Generally speaking, east and west facing walls receive the most

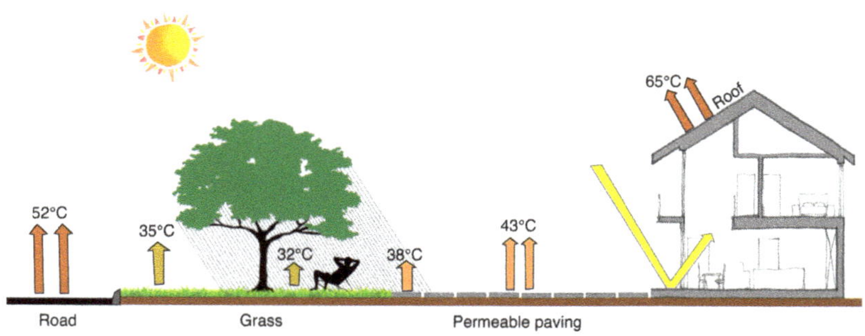

Figure 4.14. Indicative temperatures of different surfaces around buildings. Temperatures are indicative of a sunny day in a warm dry climate.

radiation on the vertical surface due to the low solar angles at the beginning and end of the day. Keeping east/west facing walls shorter with few openings is beneficial in preventing heat from entering the building. Openings on west facing façades can be particularly problematic because they receive solar radiation during the warmest part of the day (the land surface has had time to warm) and the low angle sun can penetrate into a building, causing substantial overheating. As the climate changes, there is a need to consider how best to alter the design and construction of our buildings to ensure that they continue to provide shelter and thermal comfort to their occupants. Luckily, by comparing projections of future climates with those found already elsewhere around the globe, this does not need to be a process of trial and error.

4.6 Learning from other architectures

Civilisation has evolved over the last few millennia during a period of relatively constant climate allowing primitive architecture, defined as buildings of preliterate societies, whose knowledge of architecture and construction comes from word-of-mouth, to attain high levels of performance against environmental stimuli. This implies knowledge of the performance of building materials and the effects of climate. Vernacular architecture, referring to architecture native to a particular context or location, which has been informed by factors such as tradition, environment and social norms, can be considered as the evolved form of primitive architecture. Hence, vernacular architecture, urban layout and culture (the way we use buildings) have evolved to be resilient to local climate conditions, to provide both shelter and thermal comfort. For example, architecture at lower latitudes have evolved to minimise the urban heat island (discussed further in chapter 7) and to prevent buildings overheating under intense sunlight. Likewise, buildings in the tropics have to be resistant to high winds and driving monsoon rains. Climate change is set to alter the status quo, changing the way our buildings will need to perform. By examining what architectural responses have evolved elsewhere, we can draw inspiration of how to adapt our building designs to the impacts of climate change.

The shape of a building has historically been informed by the local climate of the region in question, either to keep heat in or to keep heat out. Victor Olgyay in his book *Design with Climate* expresses the opinion that the optimum shape for a building is one elongated in an east–west direction. This allows bot only minimum solar receipt in summer on the east–west façades but greater solar access in winter on the south façade (in Northern hemisphere). In hot, arid regions, however, there is still a great risk of overheating in summer, so the building becomes more inward looking, with interior courtyards. Courtyards have been a dominant feature in Middle Eastern architecture for centuries. In the town of Ur (Iraq), excavated foundations have shown that courtyards were incorporated into building design as far back as 2000BC. It seems then that humans have been using architectural responses to high temperatures for millennia. Over time, these have migrated along with people to other locations where they have been adapted to different climates and cultures.

The south of Spain is often thought of as a haven for European tourists along the Costa del Sol; however, north of the coast are mountain ranges and plains, which receive the most intense sun of anywhere in Europe. For the majority of the year, these plains enjoy a relatively mild climate but during the summer temperatures can regularly exceed 40 °C. As such, the local architecture over the millennia has evolved to not only keep this intense heat out and safeguard the building occupants but also to provide thermal comfort during the hottest periods of the year. Lessons in the application of heat sensitive design can be found in the palaces of the region. The Alhambra located in Granada is a complex of both palaces and fortresses dating originally from 889 as a small fortress built upon Roman fortifications. It was converted to a royal palace in 1333 for Yusuf I, Sultan of Granada. However, after the Christian Reconquista in 1492, ending ~770 years of Islamic rule of the Iberian Peninsula, the Alhambra became the site of the Royal Court of Ferdinand and Isabella and was updated to better reflect the current Renaissance styles. It was further updated and added to by Charles V, Holy Roman Emperor, who added a further Renaissance palace. As such, the Alhambra and other regional examples such as the Royal Palace of Seville are a fusion of Romanesque and Islamic architectural styles, termed Mudéjar, which flourished in the period following the defeat of the Moors and the return of Christian rule to this region of Southern Spain. The Mudéjar style did not involve the creation of new shapes of building like the Gothic movement, but rather an incorporation of Islamic art and architecture into the building forms of existing medieval Christian architecture.

Central to the palaces which make up the Alhambra are decorative courtyards such as the Court of the Lions (figure 4.15). Courtyards are an extension of interior rooms, accessed by covered walkways, defined by the differentiation of three different spaces: the exposed centre, the transitional edge and the protected building interior. Of these spaces, the transitional space offers an attractive living space during the heat of the day, well-lit by diffuse and reflected sunlight but shaded from the direct heat of the Sun. Unlike the building exterior which contains few openings, the walls bordering the covered walkways and galleries contain many openings onto the courtyard beyond. The centre of the courtyard is not intended as an occupied space as might be expected in courtyard gardens in the UK, instead this is home only to plants and fountains. The interior of the building beyond the covered walkways and galleries above, surrounding the courtyard are cool and dark. This is a significant departure from the way that we currently use buildings in the UK, where we expect our building interiors to be well lit.

Shading is the most effective defence against the heat of the Sun and should be considered as the first architectural response to keep a building cool. Shade is integral to the success of a courtyard. During the middle of the day when the Sun is highest, the extended eaves, covering external galleries and walkways protect the interior floors and walls from direct solar radiation. Since the height of the courtyard is typically greater than it is wide in plan, there is shading at the beginning and end of the day as the external walls shade the courtyard's exposed centre. Courtyards are often augmented with water (ponds or fountains) and vegetation to provide cooling from evapotranspiration. This can be further enhanced by spraying water on the

Figure 4.15. The Court of the Lions within the Alhambra (Granada), note the fountain as the central feature of the courtyard and the extended eaves and covered walkways surrounding the courtyard. This [Alhambra-Granada-2003] image has been obtained by the author from the Wikimedia website where it was made available by [comakut] under a CC BY-SA 2.0 licence. It is included within this article on that basis. It is attributed to [comakut]. https://en.wikipedia.org/wiki/File:Alhambra-Granada-2003.jpg

courtyard floor a few times a day, which when it evaporates cools the surrounding air and building fabric. In dry climates, this also raises the humidity to a more comfortable level. The Court of the Lions contains not only a central fountain but also a series of channels to guide water throughout the courtyard and into the covered spaces at the edge (figure 4.16). This increases the evaporation of water within this enclosed space, cooling the air trapped in the courtyard, enhancing the contained microclimate, but also cools the high thermal mass floor coverings, reducing the radiant temperature.

Another important aspect of the Alhambra are the gardens and the extensive use of water throughout the landscape. Water and gardens have a special role in keeping surrounding buildings cool. Trees provide shade reducing temperatures of nearby surfaces (figure 4.14), hence in larger courtyards trees are used to help keep the environment cool and shade the courtyard floor helping to reinforce the microclimate effects. But water is perhaps the most important feature—to evaporate water takes a lot of energy, and hence the evaporation of water is effective at cooling the local air and nearby surfaces. As stated above, water is a prominent feature in the courtyards at the Alhambra (and in courtyards elsewhere), but the Alhambra also shows other uses for water around buildings. The pool shown in figure 4.17 serves several purposes—it is decorative, it cools the air via evaporation and it can also introduce light reflected off the surface into the rooms beyond. In a warm dry

Figure 4.16. The Court of the Lions within the Alhambra, viewed from beneath the covered walkways, note the channels to guide water throughout the courtyard and additional fountains to wet the floor.

climate, the evaporation of water also increases the level of humidity to a more comfortable level.

The Alhambra being a collection or royal palaces, gardens and fortresses is justifiably grand. The measures employed throughout the Alhambra are a result of the wealth of the occupants and the Mudéjar architectural style, these will not all be applicable to a dense urban setting or different locations. However, we can see variations on the responses showcased in the Alhambra across many other locations.

Across the Mediterranean countries, the interior of buildings are kept cool during the day as they hold onto the cool of the night due to the high thermal mass of the walls and floors. At night or in early morning, the interiors are opened up and exposed to the cool air outside. After purging any stored heat from the day before, they are shut up again to keep direct light and warm air out during the day. This is especially important for upper floors, the bedrooms, which must be kept cooler than other living quarters in order to achieve thermal comfort suitable for sleep. Homes in many European countries employ mechanisms to keep the heat of the Sun out of interior spaces during the day, through use of shutters, curtains, blinds, awnings or some other means such as plants (examples of these can be seen in figures 4.18–4.20).

The city of Valencia (Spain) offers further inspiration as to how to keep heat out of a modern city. The Old City of Valencia shows influences from the Middle East with compact stone buildings with few openings. However, as the city has expanded over time, buildings have gotten larger with more windows for increased ventilation. Figure 4.19 shows two images of the more modern Valencia, with tall apartment blocks lining narrow streets. The height of these blocks is not purely a result of land prices—the height is matched to the street width so that the street below is

Figure 4.17. The Portico and nearby pool at the Alhambra. This [Portico and pool Alhambra] image has been obtained by the author from the Wikimedia website where it was made available by [Sebastian Appelt] under a CC BY-SA 4.0 licence. It is included within this article on that basis. It is attributed to [Sebastian Appelt]. https://commons.wikimedia.org/wiki/File:Portico_and_pool_Alhambra.jpg

perpetually in shade and buildings mutually shade each other. Despite this, all openings on the façades still have shutters, blinds or awnings.

Another common site within Valencia is the prevalence of roof terraces and gardens (figure 4.20). These outside spaces provide a useful additional living space for early morning or in the evening, when it is not too hot. Vegetation or awnings located on this terrace will shade the roof below and prevent heat from solar radiation entering the rooms below.

As we move around the Mediterranean we can see other examples of regional responses that have evolved to protect building occupants from the heat of the day. Santorini is the southernmost island of the Cyclades, a group of arid islands in the Aegean Sea between mainland Greece and the island of Crete. It is a popular tourist location,

Figure 4.18. Houses in Cannes (France) with closed shutters to keep out the heat of the day.

Figure 4.19. Streets in Valencia (Spain), tall buildings frame narrow streets providing mutual shading.

largely due to its idyllic Mediterranean whitewashed architecture. The island of Santorini contains many examples of the Cycladic architectural style (figure 4.21). The many domed and vaulted roofs have a twofold purpose: to shed water and heat effectively. Santorini receives little rainfall, which falls primarily during the winter months and rainwater harvesting has been prevalent on the island for many centuries. All roofs are designed to shed water in some fashion, the vaulted roofs shed water very rapidly before it

Figure 4.20. Roof terraces in Valencia, providing an external living space for early morning and evenings.

Figure 4.21. Domed and vaulted roofs in Imerovigil on the island of Santorini (Greece). This [Imerovigli 02] image has been obtained by the author from the Wikimedia website where it was made available by [Berard Gagnon] under a CC BY-SA 3.0 licence. It is included within this article on that basis. It is attributed to [Berard Gagnon]. https://commons.wikimedia.org/wiki/File:Imerovigli_02.jpg

can evaporate, so that it can be stored in cisterns beneath the buildings. The curved roofs made of a high thermal mass material such as concrete also help keep the buildings cool. In hot sunny locations, intense solar radiation heats the building fabric, warming the interior spaces over the course of the day. The curved surfaces help limit this warming by insuring that only a small area is directly facing the Sun at any given time, thereby reducing the intensity (W m^{-2}) of impinging solar radiation. As the external temperatures fall at the end of the day, the vaulted and domes roofs have increased surface area compared to a flat or pitched roof, over which to radiate heat away out to space. This increased radiative heat loss cools the internal spaces in the evening, improving the thermal comfort of the occupants and purges the building fabric of heat accumulated during the day so that the cycle can begin again the next day.

So far, we have considered the architectural responses found in warm dry climates and in warm maritime climates. If we go further south, there are further lessons we can learn from the tropics. The warm humid climate, potentially the most uncomfortable or humans, emphasises the need for shade, reduction of solar radiation on the east and west façades but also the need to provide access for wind flow. This results in not only an elongated east–west form but also a large roof overhangs and vegetation to provide shade.

Even quite light vegetative cover can have a beneficial effect in hot dry climate zones. Vegetation provides protection against glare, dust, erosion and assists in reducing air pollution. In tropical climates, however, vegetation plays a further important role, acting as a windbreak to protect buildings from high winds and driving rain. The use of vegetation as protection from the elements allows building façades to be very permeable with many openings (many without glazing). If we look at the islands that once formed British Malaya, it interesting to see how the local architecture has influenced traditional British architecture to produce a colonial style. Figure 4.22 shows Burkhill Hall located within the botanical gardens of Singapore. Burkhill Hall is considered as the forerunner to the Black and White

Figure 4.22. Burkhill Hall (Singapore), note the tall vegetation surrounding the building, the exaggerated roof overhang and the awning to further protect the balcony and rooms beyond.

colonial plantation houses that can be found across the region. Burkhill Hall sports a large roof overhang, to not only shade the walls but to also protect them from driving rain. The roof coupled with the tall vegetation means the building is well protected from intense sunlight, heavy rainfall and high winds. The building is surprising open with a latticework entrance and in the photo we can see straight through the building to the gardens beyond. The upper floor does not even have a barrier between the interior rooms and the balcony.

Building materials are typically lightweight in this climate to allow the building to be cooled rapidly via ventilation. It is these features of sheltering and openness that characterises the buildings of the tropics, not just dwellings but commercial and industrial buildings as well. Another more recent example can be found in a clothing factory in Sri Lanka. The MAS Intimates Thurulie facility claims to be the first clothing factory to be powered solely by carbon neutral energy sources and has obtained one of the highest sustainability accolades (the LEED Platinum award). The factory was designed to harmonise with its site, with indoor and outdoor spaces integrated into a park. The site plan for the facility [4], shows rectangular factory buildings orientated with the east and west façades being the shortest. While communal areas are located on the north side of the complex such as the cafeteria, which is open on one side (the north side) to the lake and park beyond. The lake and vegetation cool the air via evapotranspiration, which can then pass into the cafeteria and factory spaces, improving thermal comfort and worker productivity.

As part of its sustainability credentials, the factory is built from local soil, furnished and finished with indigenous materials such as bamboo. The factory complex incorporates traditional Sri Lankan architectural elements, such as large roof overhangs for shade and to protect from intense rainfall, courtyards and covered walkways. In this way, there is a definitive link between the building, its occupants and the surrounding landscape, which enhances not only the aesthetic but also the environmental conditions.

4.7 Summary

In this chapter we have considered the effect of a warming climate on building occupants, considered what aspects of building design are available to use to help control internal conditions and looked at what architectural responses have evolved in other climate zones. Consideration of how humans have learnt over time to live in different climates will be of great help when considering how to adapt new and existing building to a changing climate. A summary of regional responses can be found below:

Warm maritime—e.g. Mediterranean	
Climate	Summers are warm to hot with little to no rain, winters are cool to cold with moderate rainfall. Intense solar radiation, especially in summer. Variation in temperatures, large in continental locations, small in coastal/marine locations.

Layout and form	In continental locations the response is similar to hot dry zones, with compact buildings and urban areas for mutual shading, while more marine climates buildings are spread out more to allow cooling sea breezes to permeate between buildings and convect heat away. Buildings are preferably one room deep in marine areas to allow for good airflow. Courtyards are used for the same effect in continental areas and also provide protection from cold winter winds.
Structure	For continental locations, see hot dry zones, for maritime climates thermal mass is not as important because the sea provides thermal regulation; however, floors and roofs are typically heavyweight to store heat from the Sun in winter. Roofs are designed to shed intense rainfall. All surfaces are light in colour to reflect solar radiation.
Hot dry—e.g. Middle East	
Climate	Lots of intense solar radiation, leading to large ranges of diurnal and annual temperatures. Low humidity and low precipitation. Need to consider glare off of surfaces.
Layout and form	Buildings are compact and located close to each other for mutual shading. Buildings are enclosed and inward looking with courtyards and few external openings. Larger buildings are cuboid shapes to minimise surface area and so limit solar receipt. Spaces near east and west façades are non-living spaces to provide a thermal barrier against the low sun angles. Rooms are deep, covered walkways used to prevent heat being reflected or radiated in to rooms. Fountains or water features are used to cool patios and courtyards. Flat roofs or Domed roofs are used to limit solar receipt and maximise radiative heat loss.
Structure	Windows are relatively small, particularly on external walls, and are shaded from direct solar radiation by shutters or shades. Walls are typically thick and made of thermally massive materials. The roof is solid with a reflective upper surface, sometimes a ventilated double roof is employed with the upper roof shading the roof below in contact with interior spaces. Surfaces are light in colour, sometimes with whitewash to maximise reflectivity.
Warm humid—e.g. Sri Lanka	
Climate	Lots of rainfall with high humidity. Smaller diurnal and annual range of temperatures with a high even temperature all year round. Generally, only low winds speeds. Intense solar radiation due to latitude, with high diffuse element rom humidity leading to high sky glare. Intense rainfall and often electrical storms in the afternoon.
Layout and form	Buildings are elongated in the east–west axis to minimise low solar angle sunlight. Urban layouts are sparse to allow what little wind there is to penetrate between buildings. Buildings are preferably

(*Continued*)

	only one room deep opening onto verandas, galleries or balconies to allow for better ventilation penetration. Buildings occasionally built on stilts to access greater wind speeds away from the ground. Living areas are typically open plan to allow for better ventilation. Externally, large roofs are pitched to efficiently shed rainfall with large overhangs to protect against sky glare and protect the building façades from intense rainfall and sunlight, tall vegetation also used to shelter the building.
Structure	Openings are large to allow for adequate ventilation. Sections of façade are openable, either by folding walls, or louvered openings (Jalousie windows), other sections of façade may be permeable screens or latticework to provide privacy but still allow airflow. Materials are typically lightweight with low thermal mass to allow rapid cooling from airflow. Double skin ventilated roofs often used to limit solar receipt through the large roof. Buildings are generally light in colour to increase reflectivity but the humid climate leads to algae and fungus growth, limiting the effectiveness of this measure.

References

[1] Roelofsen P 2002 The impact of office environments on employee performance: the design of the workplace as a strategy for productivity enhancement *J. Facil. Manag.* **1** 247–64

[2] Liu C, Kershaw T, Fosas D, Ramallo Gonzalez A P, Natarajan S and Coley D A 2017 High resolution mapping of overheating and mortality riskBuild. Environ. **122** 1–14

[3] Alnuaimi A, Natarajan S and Kershaw T 2022 The comfort and energy impact of overcooled buildings in warm climates *Energy Build* **260** 119–38

[4] Details of the MAS intimates thurulie facility can be found in the downloadable publication. https://lafargeholcim-foundation.org/Publications/mas-intimates-thurulie-clothing-factory-in-sri-lank

IOP Publishing

Climate Change Resilience in the Urban Environment
(Second Edition)

Tristan Kershaw

Chapter 5

The urban microclimate

For the first time in history, more than half the world's population now lives in urban areas. This is a recent phenomenon, in 1900 only 14% lived in urban areas, but this number is expected to grow to 60% by 2030 and 66% by 2050. Despite this rapid urbanisation, cities and urban areas occupy less than 2% of the Earth's land surface. To accommodate this rapid urbanisation, we will see the creation of new cities and substantial growth of existing cities over the next few decades. The greatest rates of urbanisation are anticipated to be in Africa and Asia, where some countries are expected to experience greater than a five-fold increase in urban populations by 2050. These new and expanding cities will face a series of challenges from a changing climate, such as changes in rainfall, temperature, increases in the frequency and severity of extreme weather events and potentially shifting weather patterns. This situation is made more complex because urban areas also exhibit their own microclimate arising from the presence of buildings and human activity. This chapter will explore the origins of the urban microclimate and the effects this can have on the humans who live in cities.

5.1 Introduction

It has long been known that cities generate their own microclimate and are typically warmer than their rural surroundings, this 'mesoscale' effect is known as the urban heat island (UHI). Luke Howard (who is best known for his cloud classification system) first showed the existence of this urban microclimate and measured the temperature difference between London and its surroundings to be 2 °C 200 years ago. The UHI is not constant, it varies according to weather conditions, season and the time of day. Additionally, the UHI is not constant across a whole urban area, it varies according to the size and proximity of buildings. The heat island effect is generally greatest during calm, warm weather and is most pronounced at night.

doi:10.1088/978-0-7503-5262-8ch5

Since Luke Howard measured the UHI of London in 1818, the city has grown. By the 1960s, the UHI had increased to a ~5 °C difference between the city centre and the rural surroundings and by the end of the last century the UHI had increased to ~7 °C. This temperature difference arises due to a number of physical processes, which are the subject of this chapter.

As you might expect, the UHI has a profound effect on the way buildings within cities behave, and hence on the comfort and health of the building occupants. The origins of the heat island lie in the fact that buildings, roads and pavements store heat gained during the day, both from solar radiation and from human related activity, such as traffic exhaust and heat loss from buildings. The combination of the city's geometry, which traps heat at street level, and reduced hydrodynamical cooling from hard, impermeable surfaces coupled with improved drainage to quickly carry water away before it can evaporate, alters the heat balance of urban areas compared to rural ones. Additionally, cities tend to have fewer trees and green areas compared to rural areas, resulting in less cooling from the evapotranspiration of water from plant leaves and the soil.

Urban areas have a larger surface area than rural areas over which to absorb solar radiation but also to lose heat to the atmosphere by radiation, convection, conduction and evaporation. However, the geometry of cities leads to the formation of atmospheric boundary layers, which limit the vertical transport of heat away from the cityscape. The UHI represents a partitioning of the atmosphere above urban areas exhibiting reduced vertical transport of air, caused by changes in surface roughness, temperature and humidity gradients. The UHI can be considered to be made up of three principal components:

- Urban surface layer: The heated surfaces of buildings, roads and other surfaces in the urban area.
- Urban canopy layer (UCL): This extends from the surface to the rooftops or the tops of tree canopies.
- Urban boundary layer (UBL): This extends upwards from the top of the canopy layer as a result of heat and moisture transfer between the urban area and the air above. It acts as a mixing layer between the urban environment and the surrounding atmosphere, and can be likened to a bubble of warm air sitting on top of the urban area.

5.2 Boundary layer creation

The Earth's atmosphere is defined by many different layers. The lowest level of the atmosphere is the homosphere, which extends up to ~100 km and represents the part of the atmosphere where atmospheric gases are considered fully mixed. Within the homosphere, there are four subregions (in descending order): the thermosphere, mesosphere, stratosphere and the troposphere. The troposphere extends upwards from the surface to ~10 km, and contains ~75% of the Earth's atmospheric mass and the vast majority of the planets water vapour. This layer is further partitioned into the free troposphere and the planetary boundary layer. The planetary boundary layer (PBL) is the part of the troposphere that is influenced by contact with the planetary surface. The height of the PBL depends on the strength of the atmospheric mixing generated by the surface. Typically, its depth varies according to a diurnal (24-h) and seasonal cycle. During the day, solar radiation generates strong thermal atmospheric mixing as a result of convection currents that extend the PBL up to ~1–2 km, while at night the cooling of the surface relative to the atmosphere causes a downward flux of heat supressing atmospheric mixing and the PBL contracts to <100 m. The same thermal processes contribute to greater mean PBL depth during the summer relative to winter.

Generally, a boundary layer can be considered as a layer in a fluid (i.e. a gas or liquid), where transfer processes are influenced by the properties of the underlying surface. A discussion of these processes and momentum transfer is therefore needed as a background to this chapter in order to consider different aspects of exchange between urban areas and the atmosphere above.

Figure 5.1 shows the formation of a boundary layer (red line) over a smooth flat surface immersed in a nonturbulent moving fluid. When the direction of flow is approximately parallel to the surface, the flow is said to be *laminar*. In laminar flow, momentum transfer is between in individual molecules. As the thickness of a laminar boundary layer increases, it becomes unstable and eventually breaks up to forming chaotic swirling motions, termed the *turbulent boundary layer*. A second laminar layer of restricted depth, termed the *laminar sublayer*, forms immediately above the surface and below the turbulent layer. The transition from laminar flow to turbulent flow depends on the ratio between the inertial forces associated with the flow over the surface and the viscous forces generated by the intermolecular attraction, sometimes called internal friction. The ratio between the inertial and viscous forces is called the Reynolds Number (Re), after the British scientist Osborne Reynolds who first showed that the onset of turbulence for liquid in a pipe depended upon this ratio. When Re is small, viscous forces dominate and the flow tends to remain

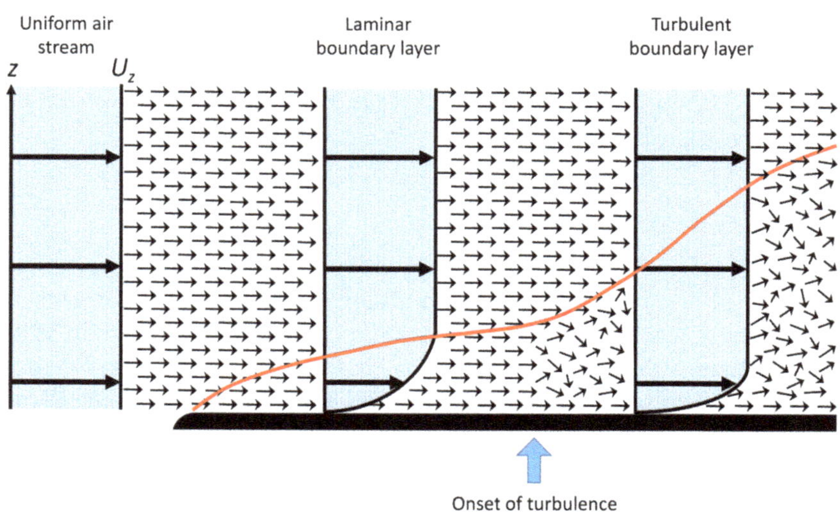

Figure 5.1. Formation of laminar and turbulent boundary layers and associated velocity profiles for a fluid flowing over a smooth plate (the vertical scale is highly exaggerated).

laminar, but when the ratio increases beyond a critical value (Re_{crit}), inertial forces dominate and the flow becomes turbulent.

The Reynolds number can be calculated from $Re = Ud/v$, where U is the free-stream velocity, d is determined from the dimensions of the system and v is the coefficient of kinematic viscosity, a property of the fluid. As a fluid flows over a flat plate, as in figure 5.1, d is the distance from the leading edge. For a very smooth plate exposed to a nonturbulent, parallel flow of air Re_{crit} is of the order of 10^6.

In order to conserve momentum across the boundary between the fluid and the plate, the fluid velocity has to be zero at the surface. For both laminar and turbulent boundary layers, the velocity increases from zero to the free-stream value U at the top of the boundary layer. The velocity profiles shown in figure 5.1 indicate that in a laminar boundary layer, the velocity increases constantly to U at the top of the boundary layer, while for the turbulent boundary layer, the velocity reaches U at the top of the laminar sublayer. Remember that velocity is a vector and not a scalar value, so direction is important. In terms of figure 5.1, the maximum velocity in direction z is achieved by the top of the laminar sublayer and above in the turbulent boundary layer this component of velocity (U_z) is maintained, but since the fluid flow is turbulent, there are other velocity components as well.

For a fluid in laminar flow over a relatively smooth surface, movement is predictable and the resultant boundary layer has well-defined profiles of velocity, vertical flux and temperature. In contrast, turbulent flow is unpredictable both spatially and temporally. We have seen in figure 5.1 how the depth of a laminar boundary layer above a flat plate increases with distance from the leading edge. In the same way, the depth of a turbulent boundary layer depends on the 'fetch' across a uniformly rough surface. Figure 5.2 illustrates how the wind profile varies as flow moves from a relatively smooth surface, such as open grassland, to a rougher surface, such as forest. The rougher surface exerts more drag than the grass and the

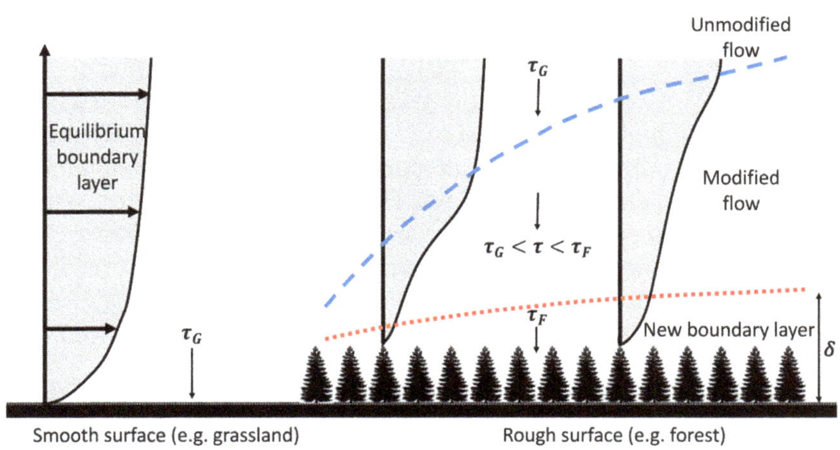

Figure 5.2. Development of a new boundary layer (dotted line) as air flows from a smooth surface to a rougher surface. The dashed line represents the boundary between the modified and unmodified flows where the vertical momentum flux is τ_G (not to scale).

air is decelerated as it moves to the rougher surface. The boundary layer that had formed over the smooth surface is disrupted by the change is roughness creating a layer of modified flow. As the flow progresses over the rougher surface, a new equilibrium boundary layer develops with depth δ, in which velocities, vertical flux (τ) and temperatures are characteristic of the rougher surface. The depth δ can be linked to the upwind fetch x. Between this new boundary layer and the unmodified flow above is a mixing region where flows are intermediate between those of the two different surfaces.

The depth of a boundary layer δ has been measured many times and, while estimates vary, the ratio between the fetch x and δ is in the range from 20:1 to 100:1. Within this new boundary layer, there are two sublayers. Very close to the rough surface (in this case a forest), this homogeneity is disrupted by the wakes of individual roughness elements (trees), leading to the formation of a *roughness sublayer*, whose properties are determined by the structure of roughness elements (e.g. foliage and plant spacing). Above this roughness sublayer is the *inertial sublayer*, within which fluxes (τ) of heat, water vapour and other gases are constant with height, provided that the flow is fully turbulent. In the inertial sublayer, which forms at $\sim 2\times$ the canopy height, the only variables of import are the friction velocity and height. This constant flux layer constitutes only the bottom $\sim 15\%$ of the atmosphere which is influenced by the nature of the surface (the PBL).

While the examples provided above relate to boundary layers forming above natural surfaces, it is easy to imagine how similar effects can be observed above cities and urban areas. This means that thanks to the geometry of our buildings and streets, which constitute the surface roughness of an urban area and hence influence friction velocities, the depths of the sublayers and vertical flux have a profound impact on the environment experienced within our cities. Boundary layers isolate the cityscape from the atmosphere above, limiting the convection of heat and pollutants away from the city.

5.3 The energetic basis and urban heat island creation

The increased surface roughness of a cityscape compared to its rural surroundings leads to the formation of a new boundary layer, partitioning the PBL into the UBL and the UCL. The UBL is a mesoscale effect, ranging from the unmodified PBL above to the roofs of the cityscape below. The properties of the UBL (vertical flux, wind velocities and temperature gradients) are influenced by the presence of an urban area at its lower boundary. The UCL, however, is a microscale effect that describes the lowest part of the PBL comprising of the air between urban roughness elements (buildings, roads, trees, etc.). The environmental conditions experienced within the UCL are dominated by the materials and morphology of the immediate surroundings and human activities. The UCL represents the anthroposphere in urban areas, it is the part of a city which humans inhabit and is consequently the part of the atmosphere that is vital for ensuring human health, wellbeing and comfort in cities.

This simplified representation of the urban climate that includes the UCL with the UBL above is blurred by the presence of air turbulence caused by wind drag and shear stresses as air flows over the buildings. As with the forest example used previously, mechanical turbulence is induced by the built-environment's surface roughness producing groups of eddies of varying scales. This creates a *roughness sublayer* that extends up to about twice the height of the buildings and includes the UCL. Above this roughness sublayer is the *inertial sublayer* which extends up to the top of the *surface layer*. The remainder of the UBL is comprised of a mixing layer where environmental conditions vary from those of the urban surface at the lower bound and extends to a height where the atmospheric properties (wind velocities, temperature gradients and vertical fluxes) are independent of the presence of the urban surface below, i.e. similar to that above the unmodified rural surface (see figures 5.2 and 5.3).

As with the global climate, the primary driver for the urban microclimate is the Sun. Urban areas differ from rural areas in that they have a lower albedo (are less reflective to solar radiation) and are constructed from high thermal mass materials. This means that considerably more solar radiation can be absorbed and stored in an urban area than in a rural one. Typically, the urban surface is slower to warm than its rural surroundings but ultimately reaches higher temperatures over the course of the day. During the day, solar radiation warms the surface (land and buildings) and the surrounding air, inducing convective instability resulting in the vertical flux of air parcels or thermals rising up into the UBL. As these thermals rise, they mix with the atmosphere above to form a boundary layer with constant temperature and almost

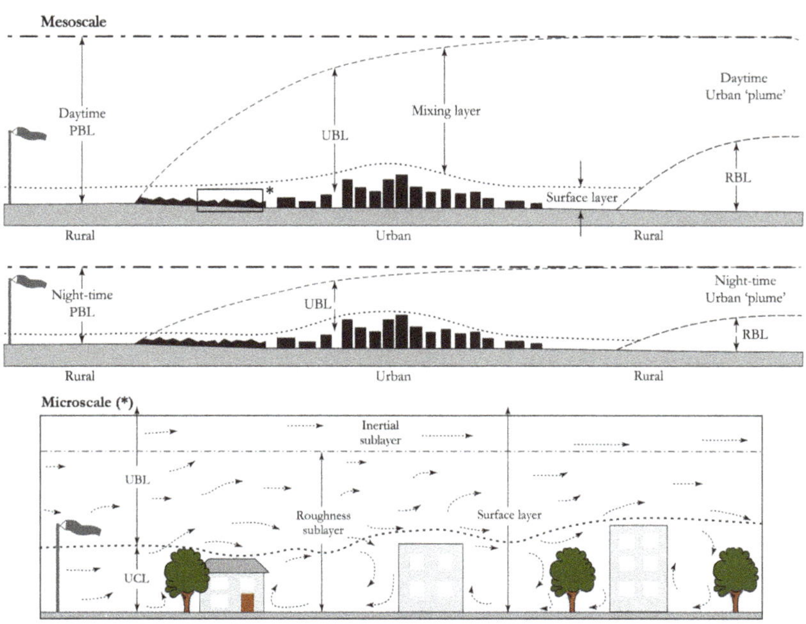

Figure 5.3. Boundary layer structure over a cityscape arising from changes in surface roughness. Top panel: daytime boundary layers; middle panel: nocturnal boundary layers; and bottom panel: close up of the microscale structures showing approximate locations of the roughness and inertial sublayers that make up the surface layer. This figure can be interpreted with the aid of table 5.1.

Table 5.1. The vertical structure of the urban climate system with typical length scales for comparison.

Layer	Definition	Dimensions	Scale
UCL	From the ground to the average height of buildings and trees.	Tens of metres	Microscale
Roughness sublayer	From the ground to 2–5× the height of buildings and trees. Includes the UCL.	Tens of metres	Microscale
Inertial sublayer	Above the roughness sublayer, with little variation of turbulent fluxes with height (<5%).	~25–250 m	Local
Surface layer	Includes the UCL, the roughness sublayer and the inertial sublayer. Flow is dominated by friction with the Earth's surface.	~300 m	Local
Mixing layer	Above the inertial sublayer, atmospheric properties are well-mixed via thermal turbulence. Usually halted by a capping inversion.	~250–2500 m	Mesoscale
UBL	The entire layer between the ground and the top of the mixed layer.	Hundreds of metres	Mesoscale

neutral stability. This buoyancy-driven process drives warmer air to the top of the boundary layer to form a stable capping inversion layer (warmer fluid above a relatively cooler fluid). This temperature inversion and stable capping layer prevents thermals rising from below, trapping heat, water vapour and pollutants released at the surface within the boundary layer. The primary difference between urban and rural boundary layers (RBL) is characterised by the intensity and depth of this capping inversion layer. The lower albedo of the cityscape compared to rural areas results in the release of more heat during the day, causing the stable capping inversion layer to be warmer and thicker than in rural areas. This boundary layer heat island is the result of the relative difference between rural and urban warming.

During the night, the surface loses the heat it had gained from solar radiation during the day to the atmosphere above. Once the surface has cooled, there is a downward flux of heat from the warm atmosphere above, causing the boundary layer to contract. Rural landscapes are quick to warm when exposed to solar radiation but are also able to purge accumulated heat quickly at night via evaporation, radiation and convection. Urban landscapes, however, are slow to warm during the day due to mutual shading from buildings and the high thermal mass of the materials present. However, once warmed they are slow to lose accumulated heat to the atmosphere. The geometry of the street's traps heat at street level. Solar radiation comprises of both direct and diffuse radiation. Diffuse radiation arises from scattering in the atmosphere and provides equal intensity on all surfaces regardless of orientation, while direct radiation depends upon the position of the Sun. However, when heat is radiated from a surface, it is entirely diffuse, being emitted in all directions equally (figure 5.4). As such, depending on the height and width of streets, heat radiated from a warm building has only a small chance of being radiated out to space, instead it may be radiated towards the ground, a

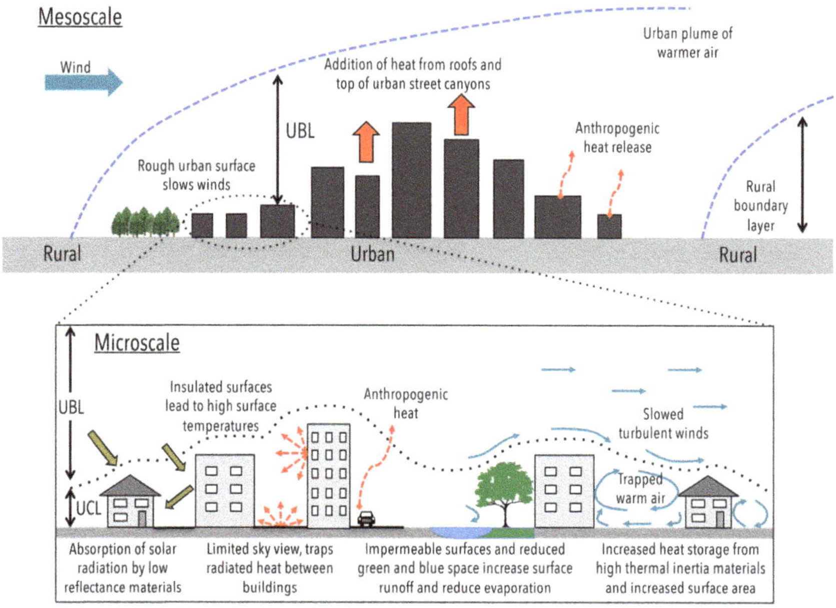

Figure 5.4. Illustration of the energetic processes that lead to the UHI, showing the UBL (top panel) and the UCL (bottom panel).

road or another building. Consequently, heat may be radiated and reabsorbed many times before it can escape from street level, an effect termed the canyon effect.

Due to this canyon effect and the high thermal mass of urban areas, urban areas continue to emit heat well into the night and continue to generate thermals, although at a lesser extent than during the day, preventing the breakup of the capping inversion layer, although it does decrease in altitude. The persistence of this capping layer, which limits heat loss from the cityscape below, means that the greatest temperature differences between rural and urban areas are often experienced during the night when rural temperatures are at their minimum.

Airflow over an urban area leads to increased wind drag and turbulence, reducing horizontal wind velocities. This frictional process is responsible for increased vertical motion above the rooftops, which forms a relatively shallow roughness sublayer above cities, extending upwards to about 2–5 × the average height of buildings. Mechanical turbulence is, therefore, the dominant driver of the UHI along with the reduced wind velocities, and both properties are strongly affected by the height and spacing of the urban roughness elements (i.e. the buildings) and the urban geometric configuration (i.e. street orientations), as such the UBL is a mesoscale effect (hundreds to thousands of metres).

The lowest part of the roughness sublayer is the UCL, which extends from the surface to around the height of the tops of trees and buildings (i.e. the height of the roughness elements). The UCL is a microscale (tens of metres) concept where the local environmental conditions are determined by the various surface properties, street orientations and height-to-width ratios (H/W) within the local urban

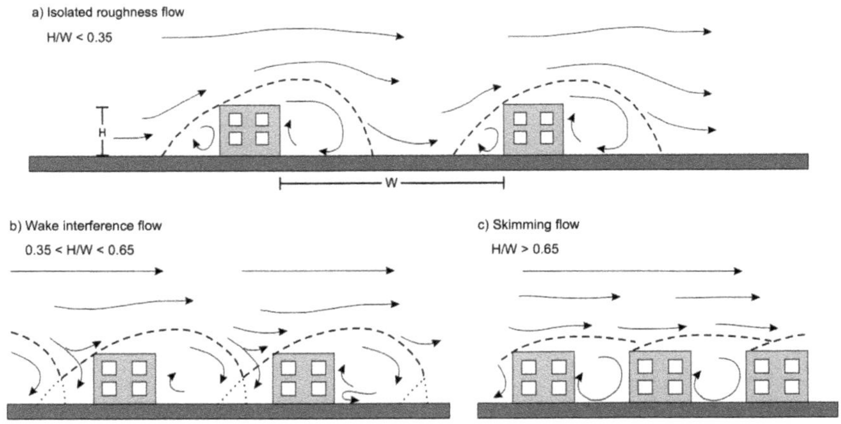

Figure 5.5. Different flow regimes above regular arrays of similar buildings at different street canyon height-to-width ratios (H/W).

environment. The flow regimes above an urban area is highly dependent on the geometric arrangement of the buildings, as shown in figure 5.5, when the canyon aspect ratio is small ($H/W < 0.35$), buildings form individual turbulent wakes, which can dissipate before impinging on the next building downstream. In this situation, the concept of an urban canopy layer is of limited use. As the urban density increases and the building separation reduces ($0.35 < H/W < 0.65$), the vortices on the leeward side start to interfere with the flow around the next roughness element downwind. As urban density increases further with taller buildings and comparatively narrower streets ($H/W > 0.65$), the vertical mixing between the street canyons and the atmosphere above becomes severely restricted due to the full development of the urban canopy layer concept, which represents the aerodynamic isolation of urban street canyons from the overlying airflow, which becomes almost laminar in manner. Under this skimming flow regime, recirculation vortices are created within the individual street canyons that reduces the mean air velocity and restricts convective energy exchange between the UCL and the atmosphere above roof level. This isolation also has the side effect that pollutants created at street level such as NO_2 and particulate matter (PM) become trapped below roof level and the vortices can create areas of very high concentration inside the urban canyons.

The UHI that is experienced in a real city is a product of not only the formation of a boundary layer arising from changes in surface roughness but also the trapping of heat due to the formation of a stable inversion layer. The addition of anthropogenic heat (from car exhausts, building heating, ventilation and cooling systems and even metabolic heat) further exacerbates temperatures. Coupled with a reduction in evaporation from improved drainage and reduced trees and ponds, which also reduces the vertical flux of air, heat and other gases (water vapour is less dense than air and rises naturally) means that the urban microclimate can be significantly different from the rural climate. A summary of the energetic processes which go into the formation of the UHI can be seen in figure 5.4.

To summarise, at a mesoscale (10^3–10^5 m) or citywide scale, the rough urban surface slows wind speeds and with the addition of anthropogenic heat, from rooftops and from the UCL, forms a boundary layer. This UBL can be likened to a bubble of warm air that sits above the urban area. As mentioned above, this may be as much as 1–2 km in height during the day reducing to only a few hundred metres at night (figure 5.3). This bubble can become distorted by the wind, forming a plume, which extends downwind of the urban area. This can cause increased air temperatures, higher pollution levels and smog in suburban or rural areas away from the main urban centre. Once we move to more rural areas the UBL gives way to the RBL, which is typified by the rural surface roughness and surface temperatures.

At a microscale (10^2–10^3 m) or building/street scale, airflow is typically turbulent, causing warm air to become trapped at street level, giving rise to the UCL. The greater the height to width ratio of the street canyons, the more isolated the air between buildings in the UCL becomes from the air above in the UBL. Airflow in the UBL is generally more laminar albeit at a reduced speed than rural areas for a given height; the rougher the urban surface (greater surface area for a given footprint), the slower the air above in the UBL. Shortwave solar radiation is reflected and absorbed by buildings, roads and other structures in the urban area according to their respective emissivities.

The geometry of an urban canyon means there is limited sky view for the longwave radiation to escape the canyon. This leads to heat being radiated back and forth between the surfaces of the street canyon, which in turn leads to elevated air temperatures depending on the humidity of the air within the canyon. This is exacerbated by anthropogenic heat from car exhausts, air conditioning systems, flues, etc. The surface roughness and the level of isolation between the UCL and the UBL, as well as rural wind speed, determine how quickly warm and polluted air in the UCL can dissipate.

5.4 Weather influence

As you might expect, the weather has a profound influence on the magnitude of any UHI experienced. Weather patterns can significantly modify the heat transfer processes between the urban surfaces and the atmosphere, with wind velocity followed by cloud cover the main variables to consider.

As mentioned in chapter 2, wind is the product of temperature variations across the face of the Earth, leading to large-scale global convection currents, which generate pressure differences resulting in atmospheric motion. At the macroscale (continental level), wind speed and direction is influenced by presence of land masses and topography which can deflect weather systems and determine wind speeds at different levels. At a mesoscale, wind speeds are influenced by the surface roughness of the urban form, which alters wind velocities with altitude and flow patterns. While at a microscale, variations in surface temperatures and wind sheltering from buildings are the dominant features. Wind velocity is the most important variable for reducing the UHI intensity—higher wind speeds (particularly over a rough surface such as a city) increase turbulence, and hence advective activity. This has the effect of disrupting the temperature inversions which intensify the UHI, resulting in a more homogenous vertical temperature profile.

After wind, cloud cover is the most important weather variable that affects the UHI because it influences how much solar radiation reaches the surface. Incoming shortwave (visible) solar radiation is absorbed and reflected by cloud tops, while heat radiated from the surface is absorbed and re-emitted by the clouds, which due to their high water content act as almost perfect black bodies in the thermal infrared part of the spectrum. Hence, increased cloud cover reduces both incident radiation (heating) and surface heat loss (cooling). The amount and type of cloud cover is important—lower altitude, thicker 'Stratus' type clouds have a greater influence than higher altitude, thinner 'Cirrus' type clouds. Thick cloud cover can reduce incident solar radiation to 10% of that on a clear day. Hence, cloud cover can have a great influence of the surface temperatures experiences in an urban area. The overall effect of cloud cover is to moderate the variation of the surface radiation balance, resulting in a smoother, flatter diurnal temperature range.

Both wind and cloud cover can be considered as representative of atmospheric stability, a measure of the vertical transport of air. High wind velocities and cloud cover are associated with increased atmospheric stability and reduced vertical transport. High wind velocities achieve this through increased advection and turbulence (termed dynamic instability), while cloud cover reduces the scope for convective thermals to carry heat and air into the high atmosphere. In contrast, calmer weather with low wind speeds and clear skies leads to a PBL that is generally considered unstable during the day and stable at night. In reality, overall atmospheric stability must be considered as being made up of different layers of varying stability. As discussed earlier in this chapter, the formation of the UHI results from convective thermals rising from an unstable surface layer, followed by a neutral mixing layer and a stable capping thermal inversion layer.

Therefore, the atmospheric stability induced by weather patterns must play a part in the strength of the UHI observed. Indeed, the greatest UHI's are observed during heatwave events, which exhibit not only lots of solar radiation and high air temperatures but also an extremely stable atmosphere. The atmospheric stability conditions above the city (stable, neutral and unstable) affect the structure and thickness of the UBL, and ultimately the temperature, velocity and turbulent flux profiles within it, which determine the vertical wind velocities at a set height. In terms of wind shear, a stable atmosphere is typified by an increased wind speed gradient with height than neutral or unstable conditions (see figure 5.6). This is due to the damping of the vertical motion of the air, which leads to a decoupling of atmospheric layers. In contrast, an unstable atmosphere has enhanced vertical motion or air which limits wind speed with height but allows greater vertical dissipation of heat and pollutants. The frequency of the occurrence of each of these different stability classes depends on the time of day, the weather, the local urban context in terms of density and surface roughness, and the heat transfer from the urban surfaces (table 5.2).

To highlight these effects and the influence of the urban area below, measurements carried out in Basel, Switzerland, showed that during the day (10:00–16:00),

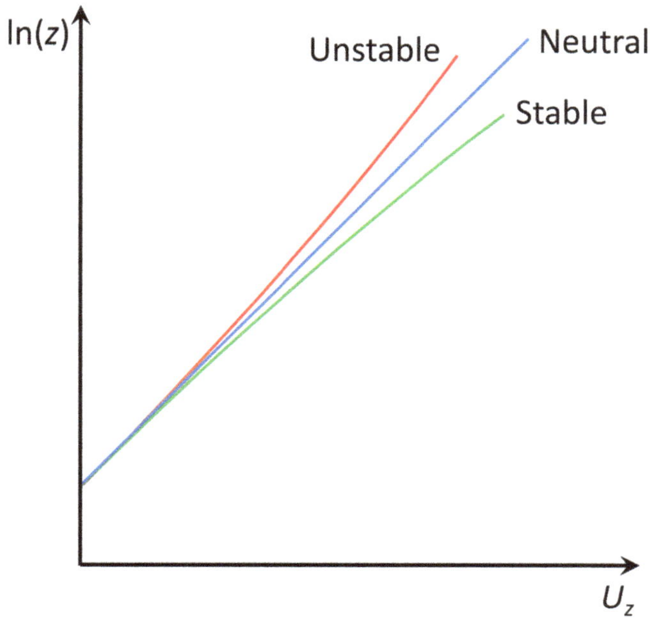

Figure 5.6. Vertical wind speed (U_z) as a function of height (z) for the three basic atmospheric stability states for airflow near the ground.

Table 5.2. Classes of dynamic atmospheric stability, detailing the processes and likely environmental conditions.

Dynamic Stability	Processes	Likely Conditions
Very unstable (free convection)	No mechanical turbulence production. Buoyant plumes produce only thermal turbulence.	Daytime, calm, no cloud cover, substantial solar irradiance.
Unstable	Thermal and mechanical turbulence production both important.	Daytime, moderate wind speed, moderate solar irradiance.
Neutral	Only mechanical turbulence. Buoyancy neither enhances nor suppresses turbulence.	Day or night-time with strong wind, overcast conditions with little to no solar irradiance.
Stable	Turbulence produced mechanically. Buoyancy suppresses vertical fluctuations and reduces turbulence.	Night-time, moderate wind.
Very stable	No turbulence created. Buoyancy suppresses vertical motion.	Night-time, calm, cloudless sky.

unstable stratification was present for more than 95% of the time above urban areas. Whereas, during the night (22:00–04:00), stable atmospheric conditions were present only 6% of the time and neutral conditions 22% of the time. Observations recorded at the same time outside the city showed that stable conditions occurred 75% over rural areas and 35% of the time over suburban areas. This indicates that the presence of the urban area, the increased surface roughness and the thermal inertia of the urban area increase the occurrence of unstable conditions, even during the night. During periods of extremely hot weather, such as heatwaves, the increased air temperatures allow a more stable atmosphere to form above cities, resulting in increased isolation of the UCL/UBL from the atmosphere above. This results in increased UHI temperatures during heatwave events and increased accumulation of air pollutants at street level due to the severely limited vertical motion of air.

Dense urban forms act as a windbreak, decreasing mean wind velocities. This reduction occurs from the transformation of the wind's mean kinetic energy into turbulent kinetic energy. Greater surface roughness from the built form and vegetation generates greater transformation of flow, this increases turbulent diffusivity, enhancing both sensible and latent flux, regardless of the vertical temperature and water vapour gradients. Strong winds promote dynamic instability and turbulent mixing, inhibiting the development of strong temperature stratification, while reduced wind velocities have the opposite effect and less heat is transferred from the surface to the atmosphere. Forced convection generated by surface roughness decreases more rapidly with altitude than the effects of thermal turbulence. These mechanical turbulence effects are therefore primarily dominant near the surface, while at greater altitudes thermal turbulence effects dominate. This is reflected in the diurnal wind profiles of the atmosphere. During the day, most of the upper boundary layer is dominated by free convection with large eddy currents resulting from thermals. At night, however, the increased stability of the atmosphere and absence of thermals, means that convective mixing in this shallower boundary layer is entirely mechanical in nature with relatively smaller eddy currents.

High-pressure anticyclonic weather systems affect temperatures in the PBL by causing air to subside and warm on top of a cooler surface air to create a subsidence inversion layer. This descending warmer air causes clouds to evaporate, thereby increasing shortwave solar radiation penetration through the atmosphere and increasing surface temperatures, and leading to atmospheric instability and vertical transport ideal for UHI formation. Conversely, low-pressure systems cause air to rise and cool to generating clouds and reducing shortwave solar radiation penetration.

The most intense UHIs have been observed during persisting high-pressure weather systems. Unfortunately, such weather systems which bring clear skies and intense solar radiation are also typical of high air temperatures. The UHI of London was measured to be as high as 9 °C during the August 2003 heatwave (compared to ~7 °C normally on clear summer days) and was likely instrumental in the increased heat mortality rate observed during this period. As the climate changes and we see warmer drier summers, the formation of high-pressure anticyclonic weather systems is likely to increase and will have an influence on future observed UHIs.

5.5 Observing the urban microclimate

Measuring or estimating the UHI can be tricky. The equations and theories that govern boundary layer physics assume an infinite, homogeneous surface. However, cities are not homogeneous! Cities have surface roughnesses, albedos and heat fluxes that can significantly change in a few tens of metres. This presents problems when trying to assess the UHI over a city: What type of measurement is best? What resolution or scale do we need? What temporal resolution do we require? The rest of this section considers the benefits and drawbacks of different methodologies to measure and estimate the UHI.

5.5.1 Field measurements and experimental studies

On-site or field measurements often require special monitoring equipment and can be labour intensive. There are two main types of on-site monitoring techniques. The first type is measurements at fixed stations, such as conventional weather stations that provide long-term weather and climate information often at an hourly resolution. The second type is shorter-term field studies, which may include fixed stations alongside additional sensors and even mobile measurement devices. Despite their relatively straightforward low-tech approach, such measurements can reveal the changes in temperature, radiant temperature, humidity and air velocity in a local area with good accuracy. In fact, assessing the microscale complexities of the urban microclimate is the greatest strength of field measurements. As discussed earlier in this chapter, the UHI was first recognised by Luke Howard in the early-1800s, who used instruments installed on his property found that London was 0.2 °C and 2 °C warmer during the day and night, respectively, than the rural surroundings. However, we need to consider that since these studies rely upon environmental measurements taken in a set location, the geometry of the local urban area needs to be taken into account and the findings may not be applicable to the wider urban area. Field studies, while expensive in terms of time and equipment, allow measurements to be made over a limited area for prolonged periods of time, potentially at a high temporal resolution (hourly or minutely). This is useful for examining the diurnal cycle of the UHI and examining how the UHI varies with season.

Even with networks of sensors spread over a city, it is difficult to overcome the site-specific nature of the individual measurements. This poses problems when trying to produce general design guidelines that are not biased by the heterogeneity of the measurement locations. It also means that data from field measurements are hard to compare because the measurements are highly dependent upon the local urban context, the geometry, materiality and the presence of any green or blue space nearby. Sadly, many studies do not include sufficient information about the urban context to make meaningful comparisons.

There are a few studies of urban scale models, typically these are idealised array of identical blocks to represent the buildings (figure 5.7). The temperature of the surfaces, the air between the blocks and at different heights in the boundary layer above can all be assessed along with air velocities to aerodynamically assess the boundary layer conditions created by the array. Alternatively, the turbulent fluxes in

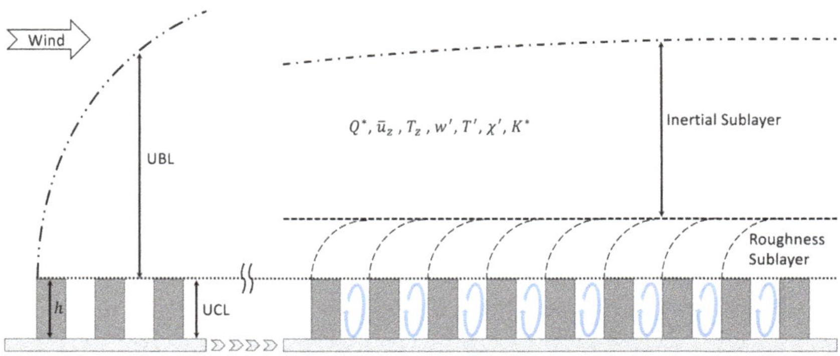

Figure 5.7. Schematic cross section through a scaled urban array providing homogeneous fetch and the resulting atmospheric structures created. Measurements are taken within the Inertial Sublayer (ISL) to characterise the boundary layer properties (symbols have been defined in the text).

the ISL over the array can be directly measured using eddy covariance techniques. Variables that may be measured include Q^* the net all-wave radiation (comprising radiated heat and the net shortwave radiation K^*), $\bar{u}_z$ the mean wind speed and T_z the air temperature at height z, also the turbulent fluctuations in the vertical component of wind velocity (w'), temperature (T') and vapour density (χ'). Such scaled arrays with these various measurements can tell us a lot about the underlying physics of boundary layer creation and how this can be modified by changes to the urban geometry (e.g. changing the H/W ratio of the streets). However, care needs to be taken when calculating the Reynolds numbers over the array to ensure that the results can be related back to a full-size equivalent. The block size and the materiality need to be carefully considered to ensure that the thermal inertia similarity is sufficient so that the scaled blocks behave thermally in the same way as real full-size buildings (i.e. that they warm and cool at a similar rate). Due to the size of the array required and a considerable amount of unobstructed upwind fetch, such scale models are typically expensive in terms of space required but do offer the ability to perform long-term studies of the impact of different environmental conditions on a boundary layer.

5.5.2 Remote sensing

Remote sensing is the process by which observations of a location can be made without physically being there. In this context, it typically refers to the collection of surface temperature data via satellite-based sensors, although field studies may also make use of portable infrared cameras. This technique allows many surfaces to be sampled at once, and is therefore considerably cheaper than individually measuring surface temperatures. The use of publicly available satellite data such as ASTER, LANDSAT or MODIS allows whole regions or cities to be considered at once. There are, however, several caveats that need to be considered when using remote sensing data to assess the UHI. The first thing to consider is that infrared cameras, whether hand-held or in orbit, do not directly measure temperature, they measure the luminosity of a surface within a particular range of wavelengths. All objects above absolute zero radiate electromagnetic radiation. The peak wavelength radiated depends upon temperature, which, as discussed in chapter 2, can be estimated by Wien's displacement law of $\lambda_{\text{peak}} = 2.9\ \text{mm}/T$. For objects like our Sun, this is in the yellow region. For objects $\sim$300 K, this is within the thermal infrared region at around 10 μm. However, if we are measuring luminosity, we need to consider the temperature and also the emissivity of the radiating surface. This requires a knowledge of the surface materials within the camera's field of view. The relatively coarse resolution of infrared cameras compared to visible cameras means that a pixel may represent the average of several surfaces and different materials. Materials such as metals have a very low emissivity, while water is almost a perfect black body at these temperatures, with concrete, vegetation, soil and glass all falling somewhere in between. Luckily, databases of surface type complied from images in the visible range allow the material type, emissivity and ultimately the temperature to be estimated with relative accuracy. Likewise, algorithms to correct for the angle of the surface relative to the camera allow temperatures of complex geometries (e.g. cities) to be measured simultaneously. This makes remote sensing the most accessible method for observing the surface UHI. Nonetheless, remote sensing measurements should be treated with caution because they only rely on the luminosity of a surface, or in other words the sensible heat being radiated at a single instant in time. While portable infrared cameras deployed in field studies can record several images over a short period of time, satellite-based sensors may only record a few images per month. As such, care needs to be given to the time of day and how this relates to the surface temperatures observed and the UHI estimated. The process of sensible to latent heat conversion (evaporation and condensation) is also neglected, which makes it difficult to assess the impact of green or blue space on the urban temperatures. There is also the issue of shadows in the urban environment, all of which may lead to an misestimation of surface temperatures and the heat balances involved.

There remains, however, a strong correlation between land-surface temperature observed by remote sensing and recorded air temperatures, particularly at night. Feng, Cai and Chapman compared remote sensing and field measurements for two similarly sized cities: Oklahoma City (USA) and Birmingham (UK) [1]. By performing a regression analysis for the two cities under different climate conditions, they found a strong correlation between the surface heat island (sUHI) and the canopy heat island (aUHI) with the relationship:

$$s\text{UHI} = m \times a\text{UHI} + b$$

where m is the slope of the regression line and b is the y-intercept at $a\text{UHI} = 0$. Table 5.3 shows the different values of m obtained under different environmental conditions. The value of b could not be determined accurately within the required confidence interval and is not presented. What we can see is that the relationship between the sUHI and the aUHI varies not only with the environmental conditions but also with the background climate between the two cities. This makes it very hard to accurately convert sUHI data from remote sensing to aUHI without first performing this analysis for the city in question.

Table 5.3. Regression values for the relationship between *s*UHI and *a*UHI under different environmental conditions of daily average wind speed (WS) and daily accumulated solar radiation (DASR). Adapted from [1] [© 2020 Royal Meteorological Society].

Criteria	City	Regression gradient *m*
WS < 3 m s^{-1} and DASR < 20 MJ m^{-2}	Oklahoma city	0.524
	Birmingham	0.492
WS < 3 m s^{-1} and DASR > 20 MJ m^{-2}	Oklahoma city	0.483
	Birmingham	0.483
WS > 3 m s^{-1} and DASR < 20 MJ m^{-2}	Oklahoma city	0.349
	Birmingham	0.279
Urban	Oklahoma city	0.231
	Birmingham	0.456
Suburban	Oklahoma city	0.383
	Birmingham	0.350

What is interesting here is that the correlation between the two types of heat island is stronger during periods of lower wind speed. What we can also see is that the two cities present very different values of *m* for the urban location but are quite similar for the suburban location. This disparity could be caused by the more complex shading and more turbulent wind from larger buildings in the urban centre affecting the values reported by the remote sensing.

Despite the obvious limitations of remote sensing, it remains a powerful tool which allows the assessment of large areas of a city to be monitored simultaneously. The public availability of the data also makes this a very attractive option for low-cost studies. However, as a method for assessing UHI magnitude, whether surface or canopy, it has a large margin of error, leading to disparities between different studies and the cooing potentials reported for different mitigation strategies. In recent years, there have been a growing number of studies[1] using remote sensing to investigate the effectiveness of green and blue spaces to cool the urban environment, most of these are located in humid subtropical climates in Asia, while a smaller number focus on the temperate climates of North America or Europe. As an example, a study of 18 parks and 197 water bodies in Beijing (China) suggests that more regular and compact geometries (round or square) have a higher cooling impact on the urban environment in terms of temperature reduction. Other studies in Beijing show that more elongated geometries can provide cooling effects up to 300 m away. While similar studies in Chongqing (China) showed the cooling potential is strongly related to the surrounding building heights and orientations, with cooling effects felt up to

[1] For further details of the various studies mentioned in this section please refer to Ampatzidis P and Kershaw T 2020 A review of the impact of blue space on the urban microclimate *Sci. Total Environ.* **730** 139068 10.1016/j.scitotenv.2020.139068

500 m away. Although fewer in number, studies in continental Europe suggest that water bodies can have a cooling effect of anywhere from 1 °C up to 10 °C. These large disparities between different studies may be due to the urban geometries and background climates, but equally it could be due to the measurement technique not being able to account for all the heat transfer mechanisms (e.g. evaporation). We should also remember that these remote sensing measurements are only reporting infrared luminosity for a single instant in time, as such the studies are highly dependent upon the time of day, while many historical images are available to build up a more complete picture, it becomes increasingly difficult to account for changes in the urban form over time, let alone the season or weather at time of the image.

5.5.3 Numerical modelling

Using computer models to simulate the urban heat island allows for comparison of real urban geometries with a range of possible variants, or purely fictious examples. Numerical models of urban climate problems can use a variety of different physics-based tools, which can take the form of more simple energy balance type models where heat losses, gains and storage within a street canyon can be represented by electrical networks of resistors and capacitors, all the way to full blown computation fluid dynamics (CFD) models of an entire city. Simulations with CFD provide a powerful alternative to observational studies, whether field or remote, because they are able to provide useful information across the whole area under study, whereas other methods only provide a limited number of measurement points in either space or time. The simulations are also run under completely controlled conditions, with the ability to isolate different parameters and make comparative studies of different permutations. In comparison to the simpler energy balance type models, CFD models can include the movement of air, heat, moisture and pollutants. Simulations are able to provide detailed information about the coupling between temperatures and air velocities, with information about the various flow fields at a much finer scale than is feasible for observational studies.

The main disadvantages are the high computational demands of high-fidelity models, although the last few decades have seen computational power increase manyfold, and the requirement for high quality data of the study area and the environmental and boundary layer conditions outside the computation domain to prime the model. This latter part is crucial because with such models, rubbish in = rubbish out, although we should note that this requirement also exists for field studies.

As the size of the study area increases, the model becomes more complex. For this reason, many large-scale studies make simplifications to the model setup to save computational time and avoid the need for high-power computing. These can include coarsening the discretisation (i.e. making grid cells larger) or reducing the dimensionality of some of the parameters. This latter part may include simplifying the heterogeneous urban geometry to a homogeneous roughness patch, i.e., an area of the domain that has all the physics parameters (albeit averaged) of the urban area, such as temperature, albedo and roughness length, but with specific geometry. This is often referred to as a 2.5D model. Similarly, areas of water or vegetation are often simplified in a similar way with their complex geometry and physical parameters generalised and averaged to reduce the model's complexity. Heat flows are also often simplified with surfaces having an initial temperature which does not vary. This is equivalent to an urban surface that can emit heat but effectively has an infinite heat capacity, so its temperature does not vary regardless of the amount of heat it receives or loses over the duration of the simulation. These simplifications, as you might expect, can reduce the confidence we can have in the results. One way to overcome this and still have manageable simulations is to nest domains, with a high-resolution area of interest nested within a parent domain that is of coarser resolution, potentially with further simplifications, which acts as a buffer modifying the universal boundary conditions (e.g. taken from a meteorological station outside the city) for use by the daughter domain.

For example, Theeuwes *et al* [2] simulated an idealised circular city in a temperate climate to examine the influence of water body distribution on citywide cooling. They used a weather research and forecasting (WRF) model to model the mesoscale energy flows coupled with a land-surface model. In contrast to other studies, this model considered how water bodies thermal properties varied between day and night. Since the model was a 2.5D approach with the cityscape and water body reduced to a surface patch with representative roughness and temperature, this involved changing the initial conditions for each simulation. This large-scale, simplified model was able to examine the vertical atmospheric structure above the lakes under conditions representative of different times of day. They found that during the morning, cooling was observed within the first 500 m vertically, with warming at higher altitudes. While in the afternoon, when the evaporation and vertical transport is stronger, cooling was found at all altitudes over the lake. A warming effect was obvious only at lower altitude during the night, which could be attributed to contraction of the UBL. This study considered very large lakes and indicated that different water body geometries behave differently at different times of the day. While several different distributions of water body were considered in this study, the percentage of water compared to the area of the city was considerable (tens of %). This makes this study somewhat unrealistic. Another study by Ampazidis *et al* [3] used a much higher resolution model covering a much smaller idealised urban area using the OpenFOAM CFD software. This simulation contained actual geometric buildings but again the water was reduced to a surface patch of fixed temperature producing water vapour. Again, this simulation showed that different shapes and sizes of water body behave different under different conditions which may be encountered over the course of a day. Of great importance was the shared observation that smaller or cooler water bodies provide less vertical motion and instead can exacerbate environmental conditions at the surface because the induced vertical motion of the water vapour is unable to break through the boundary layer and instead can promote downwind warming as warm humid air dissipates horizontally at night. It is encouraging that different types of models at very different scales can produce similar results—even if those results are somewhat troubling. Further details can be found in the next chapter.

A growing number of studies are now using coupled frameworks of multiple modules to analyse the UHI and its impacts. Software packages such as ENVImet and PALM 4U are able to modify external boundary condition, assess local climate conditions in detail using CFD and have integrated packages for assessing the impact of green or blue space, building energy use and human thermal comfort. Such coupled frameworks are powerful tools; however, there is a risk that simulations are performed in isolation and the software is treated as a black-box with little understanding of the processes between the various modules. The danger then becomes that simulations can be performed, and the results taken as read. As such, it is still important to validate models before starting to trust the results of the simulations.

5.5.4 Overview

There are a variety of ways to observe and measure the boundary layers and the resultant urban heat island over a city. Each method has its own set of benefits and drawbacks which need to be considered. Due to the variables that are being measured, care needs to be taken when comparing different studies, whether that is urban canopy layer versus urban boundary measurements or surface heat island versus atmospheric heat island. For instance, numerous remote sensing studies have tried to assess the UHI and have found that bodies of water are by far the coolest element within the urban environmental (most remote sensing studies use daytime images not nocturnal). From these studies, cooling potentials of blue space are estimated; however, when we compare these values to those estimated from other studies, we get a range of answers. For example, studies of similar ponds in Beijing, China found the average cooling potential to be 1.9 °C, 2.4 °C and 5.3 °C for numerical modelling studies, field observations and remote sensing studies, respectively. Such variance in cooling potential indicates just how much research there is still left to do to understand how to mitigate the UHI and cool the urban environment. Table 5.4 shows some of the reasons why we might expect different observations of the UHI within different parts of the urban atmosphere.

Table 5.4. Possible energy processes which may lead to a difference between observed UCL and UCL heat islands.

Urban boundary layer	Urban canopy layer
Increased absorption of solar radiation	Increased absorption of solar radiation
Anthropogenic heat release	Increased heat storage
Increased sensible heat flux from above	Anthropogenic heat release
Increased sensible heat flux from below	Reduced evapotranspiration
	Increased absorption and re-emission of longwave radiation
	Reduced longwave radiation loss
	Reduced turbulent mixing

5.6 Implications of the UHI on the built environment

The UHI alters the daily cycle of temperatures in an urban environment, both in magnitude and also shifted in time. Figure 5.8 shows the difference between recorded temperatures at a rural location and those simulated for a dense urban location, in this case the City of London[2], for a clear summer's day with an intense UHI. As we can see the effects of the urban environment lead to elevated nocturnal temperatures. However, during the day urban temperatures can be lower than in rural locations as a result of the mutual shading of buildings and the high thermal mass of the materials present in urban location, this is termed the urban cool island effect. We can also see that daytime peak temperatures are similar, this is likely due to the fact that they stable capping thermal inversion layer above the city has yet to form and so strong thermals are still rising from the urban landscape, keeping temperatures in line with the rural counterpart.

If we subtract the rural temperatures from the urban temperatures, we can extract the effect of the urban environment (figure 5.9). We can see that at night the UHI is as large as 7 °C while during the day the urban area can be as much as 2 °C cooler in the late morning.

The magnitude of the urban heat and cool islands depends on latitude (solar angle), the size of the buildings, anthropogenic heat generation in the urban area and the different factors that govern the formation of the boundary layers. The values

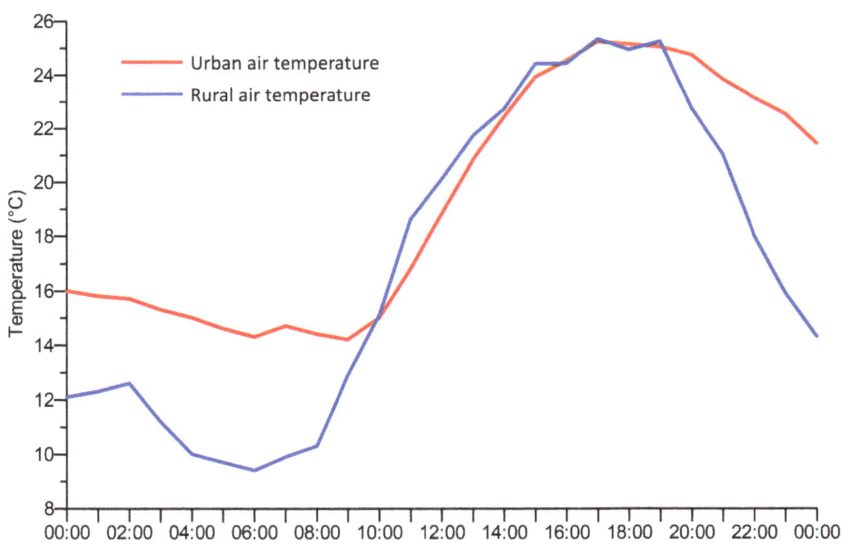

Figure 5.8. Variation between urban and rural temperatures on a clear summer's day.

[2] Data presented in this section compiled from simulations detailed in: Gunawardena K R and Kershaw T 2017 Urban climate influence on building energy use *Proc. Int. Conf. on Urban Comfort and Environmental Quality* September 2017. Genova, Italy.

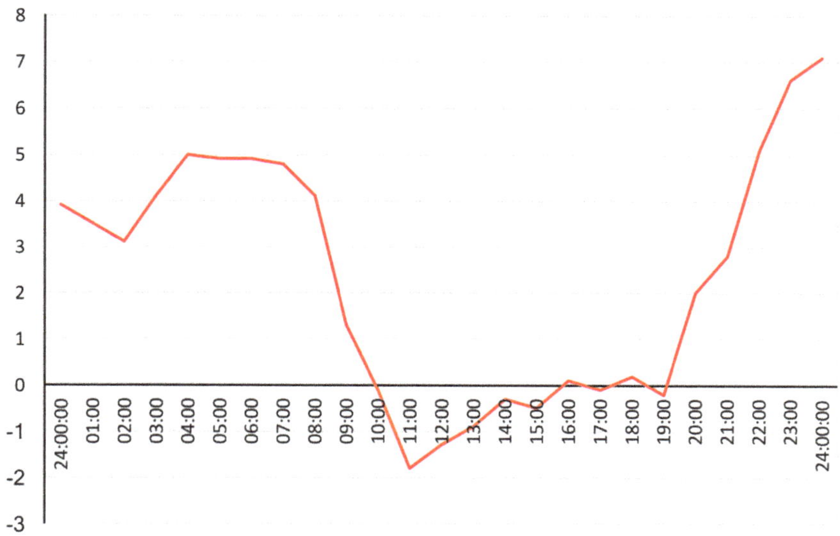

Figure 5.9. The diurnal UHI cycle, showing the difference between the urban and rural temperatures shown in figure 5.8.

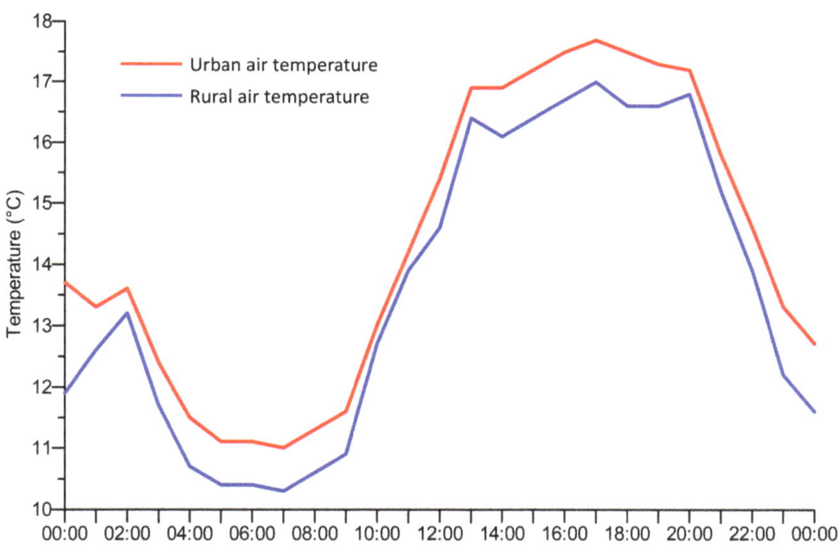

Figure 5.10. Variation between urban and rural temperatures on a completely overcast summers day.

presented in figure 5.9 are in line with those measured for the City of London. As mentioned previously the weather also has a profound impact upon heat island generation. Figure 5.10 shows recorded rural temperatures and simulated urban temperatures for a completely overcast summers day. Here, we can see no distinct heat island or cool island, instead the urban temperatures are systematically higher than the rural temperatures throughout the whole diurnal cycle. This is due to anthropogenic heat release in the urban area from car exhausts, building cooling systems and even metabolic heat. The lack of evaporation from reduced vegetation and water bodies in the urban area also limit vertical transport of heat, resulting in slightly elevated temperatures.

The UHI in cooler climates, such as that of the UK currently, can be beneficial because buildings in cities require less winter heating energy than their rural counterparts. Additionally, the cool island during the morning can move summer-time peak internal temperatures towards the end of the day and potentially outside of normal working hours. However, this means that the UHI decreases the ability of buildings to shed heat at night, significantly reducing the effectiveness of night-time cooling and this will have an impact on thermal comfort in domestic properties, which are occupied at night.

Continuing with our City of London example, we find that the presence of the UHI has an effect on the amount of energy used for heating and cooling throughout the year. Figure 5.11 shows the heating and cooling energy for a medium sized commercial office exposed to a rural climate and an urban climate. While savings are made in winter with respect to reduced heating load, this is repaid in in summer through increased cooling loads. Whether the UHI has a net positive or negative effect on the energy performance of the offices depends upon the coefficient of performance (CoP) of the heating and cooling systems employed.

The materials used in an urban area have an effect on both the temperatures experienced and their timing through their influence on the urban surface energy balance. The radiative properties of a material are its emissivity and albedo, while the energy storage properties are its mass, heat capacity and thermal conductivity. In this case, albedo refers to the amount of solar radiation (with wavelengths ~250–2500 nm) reflected by a surface. With nearly half of the solar radiation in the visible

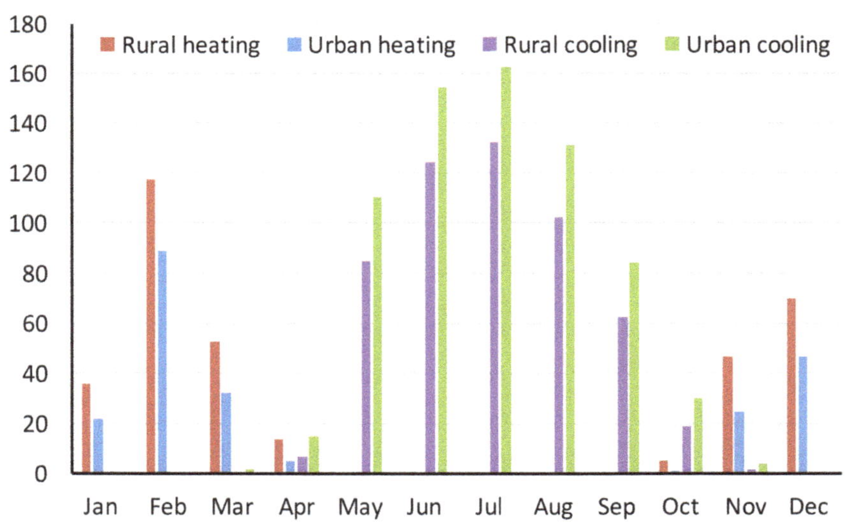

Figure 5.11. Implications of the UHI on both heating and cooling loads (MWh) for a medium sized commercial office.

region (400–700 nm), the colour of a surface is a good indication of its albedo. This is a significant factor in the determination of the surface temperatures because greater reflectivity means that for high albedo materials less energy is absorbed and stored in the material. This is where the thermal conductivity and heat capacity of the materials comes into play because they determine the thermal diffusivity of the material. The conductivity determines how quickly absorbed radiation can be transferred through the material as heat, while the heat capacity determines how much the temperature of the material will change for a given amount of absorbed energy. These determine the thermal inertia of the material, which is colloquially referred to as its thermal mass.

Heavyweight (thermally massive) materials such as stone and brick have a relatively higher thermal diffusivity, heat capacity and hence thermal inertia. This means that their temperature fluctuations over the course of a day are minimised. When solar radiation energy is incident on such surfaces, the nonreflected energy is absorbed and stored within the material, which increases the temperature of the material. As the surrounding air cools towards the end of the day, this stored heat is reradiated back to the local environment as longwave (thermal) infrared radiation. This radiated heat is diffuse in nature, and therefore is likely to be reabsorbed by other surfaces within the street canyon, contributing to the experienced temporal lag of urban temperatures compared to rural temperatures. The work of Gunawardena *et al* [4] examined the effects of different building materials on the UHI experienced, and how this affected building energy usage. It was found that lightweight materials such as timber clad or glazed façades were quick to warm up and quick to cool down, these contribute to higher daytime temperatures, and lower evening and nocturnal temperatures. Thermally lightweight materials have a lower thermal diffusivity, and hence surface temperatures increase faster and reach higher temperatures, leading to increased radiative heat loss from the building surface. Which in turn increases the mean radiant temperature outside in the street canyons and worsens daytime external thermal comfort. Meanwhile, heavyweight materials such as stone or brick demonstrated lower daytime temperatures but higher evening and nocturnal temperatures, and hence a greater maximum UHI, which typically occurs overnight. The relatively warmer surfaces of heavyweight materials overnight worsens nocturnal external thermal comfort and limits cooling of the urban area. The impact of the materiality of a building depends on it use and the timing of occupancy. For an office building which is occupied during the daytime, the use of heavyweight materials reduces peak air temperatures during the daytime and acts as a buffer limiting the effects of high internal gains within the office (e.g. from lighting, computers and people). In this way, heat can be stored during the day to be reradiated later after working hours, improving daytime thermal comfort and lowering cooling energy usage. The opposite could be considered true for residential buildings which have a predominantly nocturnal occupancy, here lightweight constructions which are quick to cool are of greater benefit, moving peak temperatures to earlier in the day and away from the occupancy period. Care does need to be taken in both cases to have adequate provision to remove heat from the building to limit overheating during the occupancy periods. The trend of retrofitting buildings with highly insulated façades and reducing the available thermal capacity has a knock-on effect on the building energy use where active cooling is used. Gunawardena *et al* [4] showed that in the context of a city centre, changing from traditional stone façades to a more insulated alternative resulted in an estimated 2.6% increase in space-conditioning load (heating and cooling). This is an example where the climate change mitigation agenda is not always beneficial and thought needs to be given to how best to implement 'energy saving' measures such as insulation. A 'light and tight' building construction is good for creating a thermally responsive environment but may not necessarily be the best solution when the urban microclimate is also taken into account. The timing of occupancy and the building use should be considered when making decisions about the materiality of buildings so as not to exacerbate the effects of climate change over the course of a building's lifetime.

In addition to the energy implications of the UHI, there are physiological implications of the UHI. During the 2003 European heatwave, which claimed up to 70 000 lives, the highest temperatures were not recorded in urban areas; however, this is where the people died. The UHI in London was measured to be as much as 9 °C during this heatwave due to the clear sunny weather. As we saw from figures 5.8 and 5.9, the heat island is a maximum at night, and herein lies the danger. This is because human physiology allows us to be exposed to high temperatures for short periods of time (hours). However, over longer time periods, the physiological strain of increased sweating, blood vessel dilation and increased core body temperature can have serious health implications. If the body is not allowed to cool overnight due to the temperature of the surroundings still being elevated, such as within a building located within an urban centre with a large UHI, the prolonged physiological stress can cause not only severe dehydration but also a variety of heat related illnesses, including heart failure.

5.7 Air quality in cities

The UHI is not just a threat to human life during hot weather and heatwaves but is also associated with poor air quality. In the UK, around 40 000 deaths per year are attributable to outdoor air pollution, and the World Health Organisation (WHO) estimate that globally over 7 m deaths per year are as a result of air pollution. Air pollution plays a role in many of the major health challenges facing the global population. Air pollution has been linked to various cancers, asthma, heart disease and strokes, as well as diabetes, mental illnesses such as dementia and increased obesity.

The health implications of air pollution accrue over a lifetime, from a baby's first few weeks in the womb, through adulthood to old age. Exposure to air pollution during pregnancy has been associated with low birth weights and premature births. Gestation, infancy and early childhood are particularly vulnerable times because the young body is developing rapidly and the heart, brain, immunological and hormone systems can all be harmed by air pollution.

Environmental variables such as temperature, humidity, precipitation, wind speed/ direction, air pressure and the mixing height (the top of the surface layer, see figure 5.3) all have important roles in determining the air quality above a city. When a high-pressure weather system sits above an area, it concentrates pollutants close to the ground, suppressing vertical diffusion. High levels of UV radiation during warm sunny weather reacts with pollutants to produce ground-level Ozone (O_3). O_3 can cause shortness of breath, coughing, wheezing, decreased lung capacity, inflammation of the airways and even death. Anthropogenic emissions such as oxides of Nitrogen (NO_x) and volatile organic compounds (VOCs) contribute to O_3 production. In addition to these health impacts, some gases act as greenhouse gases while others such as sulphur dioxide (SO_2) have a cooling effect on the local environment. Gaseous pollutants can be dispersed via horizontal wind flow and via vertical transport of air up from the surface. Rain is necessary to clear pollutants out of the air. Without rain, the concentrations of dust, smoke and other pollutants (particularly particulate pollutants) will increase over time. This presents an important side effect of the shifting of rainfall from the summer to the winter as a result of climate change. As the climate warms and summers become drier, we can expect urban air quality to worsen, and exacerbation of the associated negative health impacts.

In Luke Howard's day, air pollution could largely be attributed to the burning of coal, and the dark smoke it produced that hung over cities was highly visible. However, modern-day air pollution in the UK largely comprises of very small particulate matter, oxides of nitrogen and O_3. These are odourless and invisible to the naked eye except for an occasional brown haze from the oxides of Nitrogen that can be seen hanging over cities. Modern-day pollutants have been shown to be highly damaging to not only the lungs but also the heart and cardiovascular system. It is partly for these reasons that many cities have introduced congestion charges and several European cities are considering banning diesel vehicles by 2025 due to their production of particulate matter (PM) pollution. The following images are from the London Atmospheric Emissions Inventory (LAEI) and show the post-2013 distributions of various pollutants across the British capital (figures 5.12–5.15).

We can see from the maps that airborne pollutants are concentrated around the city centre, Heathrow airport (on the Eastern side of the city) and around major thoroughfares. The PM_{10} exceedance map (figure 5.16) shows that the consistently greatest concentrations of particulate matter are found along the major roads into and around London. This is not only due to the production of PM by motor vehicles but also the trapping of these particles within street canyons.

Increasing vegetation is often touted as a way to reduce air pollution in urban areas. While this is true for gaseous pollutants such as NO_2, which can be absorbed through stomata, the species of plant is important. Plant species with broad leaves are effective at taking up pollutants; however, these tend to be deciduous and will only have an effect during spring and summer. While evergreen trees such as conifers are active all year round, they are not as effective at taking up pollution. For PM, vegetation is not very effective at all with the deposition of PM onto vegetation being a small ($\sim$1% effect). Instead, in the absence of rainfall to remove PM from the lower atmosphere, a varied landscape with increased surface roughness is required to disperse the PM (and other pollutants). A varied landscape is aerodynamically rougher than a homogeneous one, leading to increased turbulence and vertical transport (figure 5.17). Parks and other green

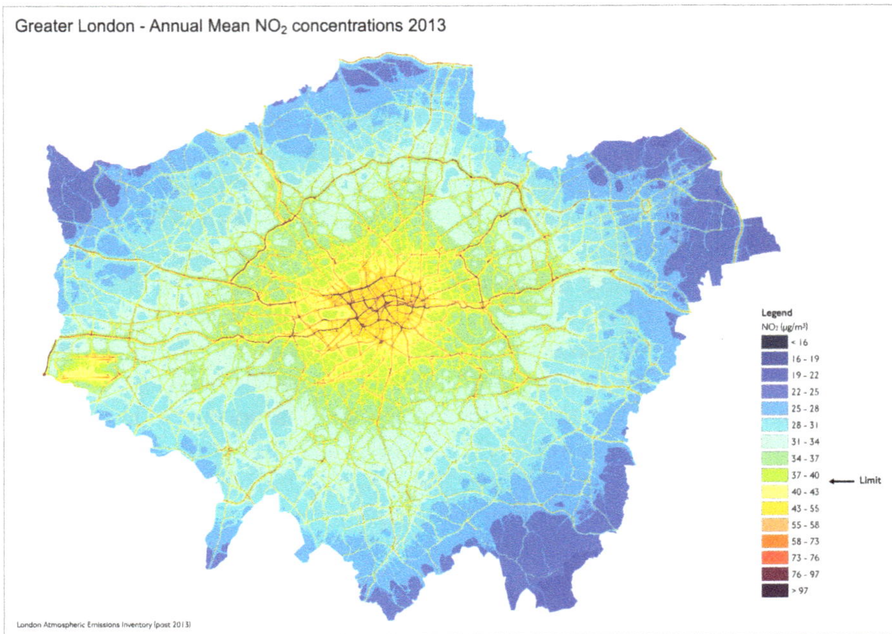

Figure 5.12. Annual mean Nitrogen Dioxide (NO_2) concentration, note how much of London exceeds the national limit for NO_2. Data from https://data.london.gov.uk/dataset/london-atmospheric-emissions-inventory-2013 (Open Government Licence v3.0).

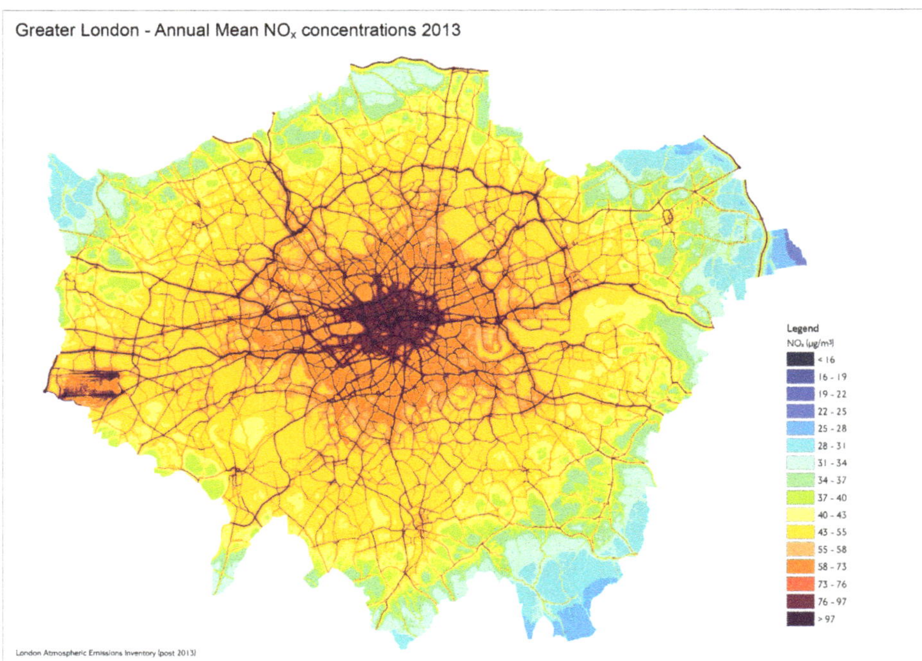

Figure 5.13. Annual mean concentration for the ensemble of Nitrogen oxides (NO_x). Data from https://data.london.gov.uk/dataset/london-atmospheric-emissions-inventory-2013 (Open Government Licence v3.0).

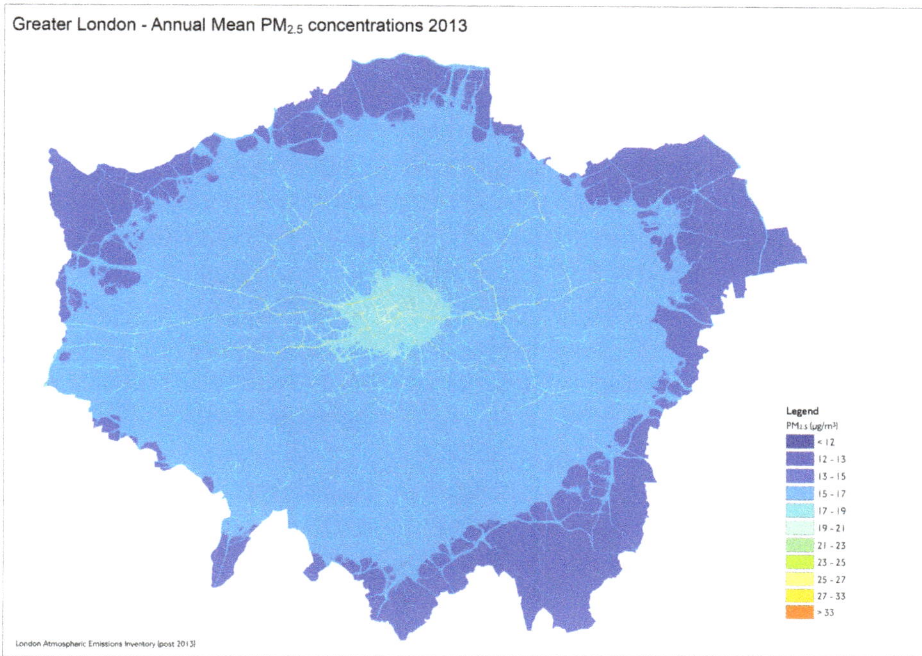

Figure 5.14. Annual mean concentration of $PM_{2.5}$, particulate matter less than 2.5 microns (µm). Data from https://data.london.gov.uk/dataset/london-atmospheric-emissions-inventory-2013 (Open Government Licence v3.0).

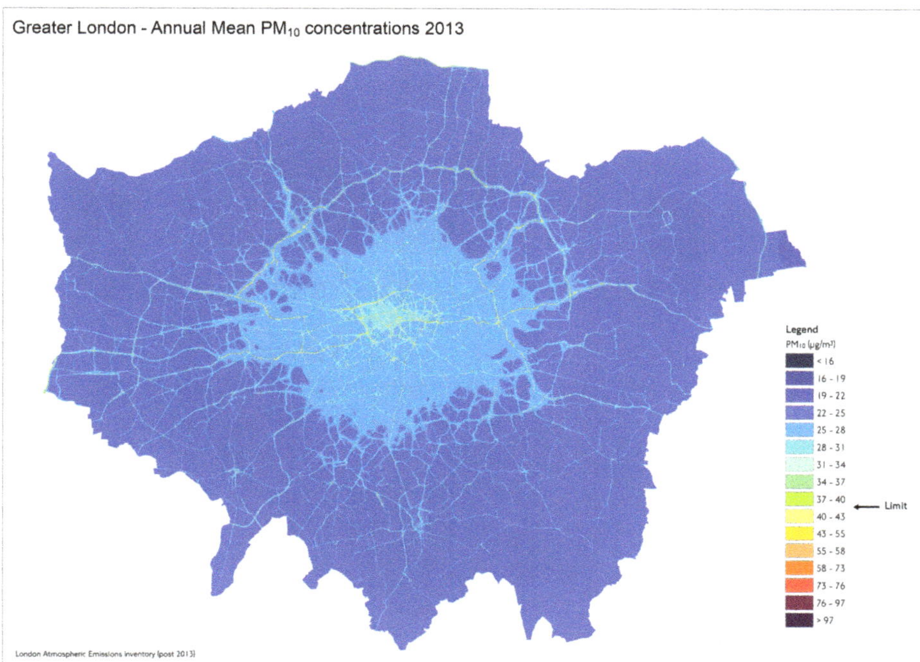

Figure 5.15. Annual mean concentration of PM_{10}, particulate matter less than 10 microns (µm). Data from https://data.london.gov.uk/dataset/london-atmospheric-emissions-inventory-2013 (Open Government Licence v3.0).

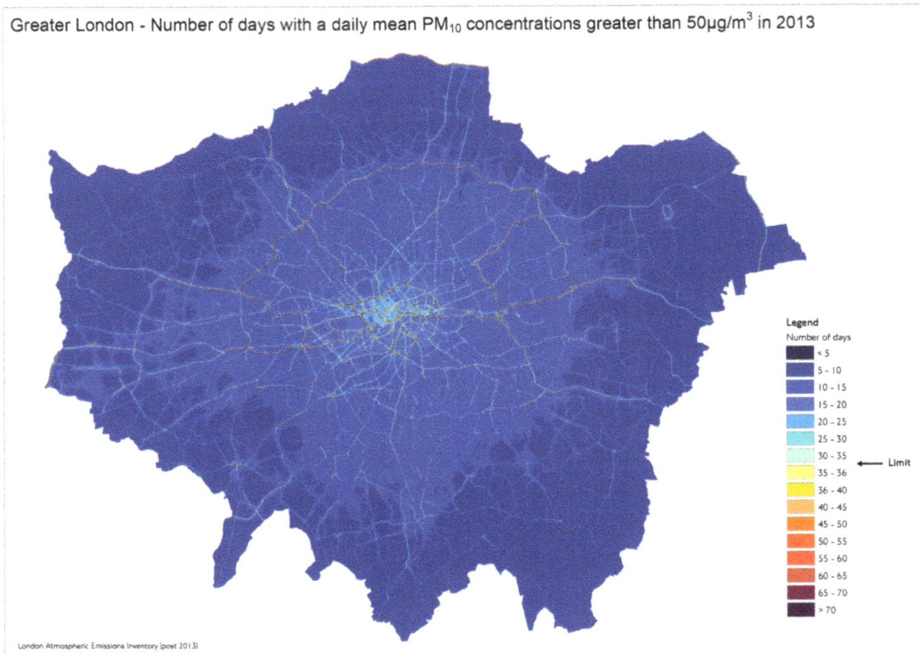

Figure 5.16. Number of days per year when the limit for PM_{10} concentration is exceeded. Data from https://data.london.gov.uk/dataset/london-atmospheric-emissions-inventory-2013 (Open Government Licence v3.0).

Figure 5.17. Parkland and buildings of different heights provide greater surface roughness to disperse pollution (top panel) while less roughness concentrates pollution at ground level (bottom panel).

spaces will reduce the concentration of air pollutants, not only through take up and deposition but also through increased surface roughness, increased evapotranspiration, and hence vertical transport (water vapour is lighter than air), and finally by providing a space into which pollutants can disperse via Brownian motion, thereby having a diluting effect on airborne pollution.

In the same way as green spaces can dilute air pollution, the urban geometry and trees lining roads or pedestrian walkways can alleviate air pollution. In figure 5.18, we are presented with two situations. In both cases, the air within the street canyon is isolated from the air above the UCL; however, in the top case the pedestrians are separate from traffic and the trees will act to filter the air within the street canyon. In the bottom case, however, the presence of the trees act to further isolate air in the canyon from the atmosphere above and air pollution can become trapped and concentrated beneath the tree canopies, thereby accentuating the risk to humans.

The challenges of safeguarding humans against high temperatures and air pollution in urban areas are inextricably linked via the environmental conditions imposed by the urban heat island. This, however, means that these issues have common solutions related to urban planning and the aerodynamic roughness of the cityscape. In addition, the presence of parks not only dilutes air pollution but also provides evapotranspiration and promotes vertical transport of air.

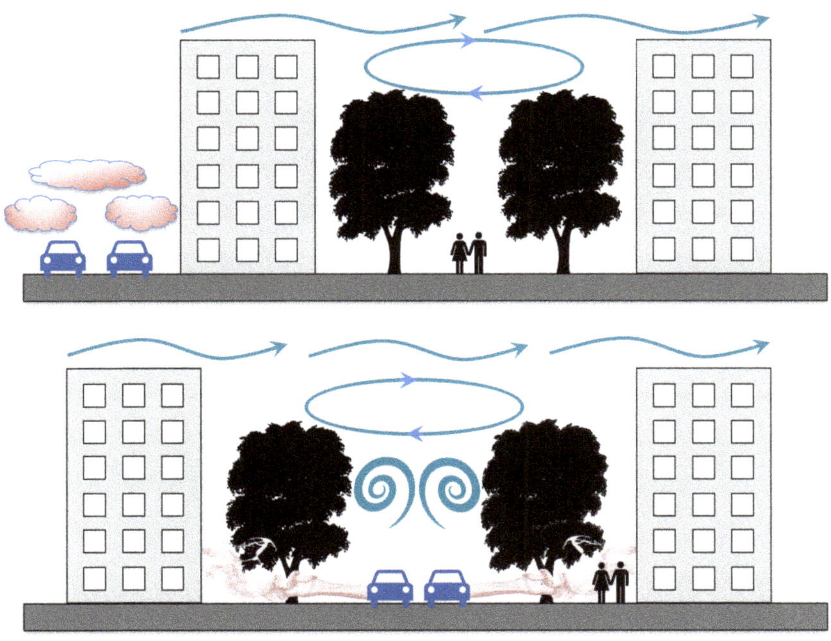

Figure 5.18. Trees and a street canyon produce a filtered avenue (top panel) with air cleaner than on a regional scale. While an avenue with traffic (bottom panel) can retain pollution at pedestrian level.

References

[1] Feng J-L, Cai X-M and Chapman L 2021 A tale of two cities: the influence of urban meteorological network design on the nocturnal surface versus canopy heat island relationship in Oklahoma City, OK, and Birmingham, UK *Int. J. Climatol.* **41** E445–62

[2] Theeuwes N E, Solcerová A and Steeneveld G J 2013 Modeling the influence of open water surfaces on the summertime temperature and thermal comfort in the city *J. Geophys. Res. Atmos.* **118** 8881–96

[3] Ampatzidis P, Cintolesi C and Kershaw T 2023 Impact of blue space geometry on urban heat island mitigation climate *Climate* **11** 28

[4] Gunawardena K, Kershaw T and Steemers K 2019 Simulation pathway for estimating heat island influence on urban/suburban building space-conditioning loads and response to facade material changes *Build. Environ.* **150** 195

IOP Publishing

Climate Change Resilience in the Urban Environment
(Second Edition)

Tristan Kershaw

Chapter 6

Planning for urban resilience

The concepts of climate change resilience and the urban microclimate are inherently linked in my mind. The world is urbanising with an increasing number of people moving to ever growing cities. If we can control the urban microclimate, we can offset the warming due to climate change and we can also reduce the energy use in cities for things like cooling. This chapter considers if we can alter cityscapes to mitigate the Urban Heat Island (UHI), make cities more comfortable and healthier places to live, and offset some of the effects of climate change.

6.1 Are cities efficient?

We are living in what is being termed the Anthropocene. The Anthropocene is an unofficial unit of geologic time (like the Jurassic or Devonian periods), which is used to describe the recent period of the Earth's history where humans (hence Anthro-) have had an impact on the planet's ecosystems and climate. This term is often used in the context of climate change, and hence most people would start the Anthropocene a few hundred years ago at the start of the Industrial Revolution, when industrialisation started emitting ever-increasing amounts of CO_2 into the atmosphere. However, there are alternative arguments that would make the Anthropocene period much longer. The earliest urban conglomerations, the world's first cites, are found millennia before most people expect. A popular example is Çatalhöyük in Anatolia (now Turkey) about 8000 years ago. These proto-cities, derived from combinations of trading networks, pre-date agriculture and it can even be hypothesised that these original cities were the reason for the development of agriculture. As cities grew, the old methods of hunter—gatherer became insufficient to meet the needs of the increasing population density. Arguments for an early Anthropocene point to greenhouse gas anomalies at about 8000 years ago and another at ~5000 years ago, both being attributed to early agriculture. The increase in CO_2 8000 years ago is thought to be as a result of land-use

doi:10.1088/978-0-7503-5262-8ch6

change, with large-scale deforestation to clear land for farming. The greenhouse gas anomaly at 5000 years ago is attributed to CH_4 (methane) from the development of wetland rice farming in Asia through a dry period with significant levels of rotting vegetation. In both of these cases, the increases in greenhouse gases are caused by human intervention, modifying ecosystems and land use. The reason for this increase in agriculture can be laid at the feet of our early cities.

Cities are often touted as being resource and infrastructure efficient, but are they? Cities are inherently demanding—they grow and prosper through the ever increasing consumption of the city dwellers. As the first cities required the development of agriculture in order to meet the needs of their citizens, we are now in a situation where the entire planet is being marshalled to fuel urban development. Growing urban demand is a millennia old problem, with a more recent acceleration due to increasing urban migration. Thus, the Anthropocene goes back thousands of years to the formation of our earliest cities and their need to sustain the population therein, requiring the clearance of land, the development of agriculture to grow food and more recently to provide resources and energy for consumption within the city.

We can consider cities as a measure of socioeconomic development—as economies grow, so do our cities, becoming more and more demanding of resources. From the large-scale clearance of land to provide food, building materials and fuel/energy, urban development has been instrumental in modifying the global climate. Demand has increased many-fold as Western consumerism has spread across the globe, this coupled with China and India's rapid growth in urban population means that the demand of cities is such that we are at risk of terminal consumption, where our need for resources will consume the very ecosystem that we inhabit.

6.2 The garden city movement

Following the Industrial Revolution, many cities across Europe faced a dramatic increase in population growth, this was combined with a migration of people from rural areas to urban areas to obtain work in the new industrial sectors. This is a trend that continues today as countries in the Global South undergo socioeconomic change and begin to rapidly urbanise. Although cities were inviting to many due to the opportunities they offered, problems such as pollution and the growth of informal settlements began to appear. These growing settlements had a much high population density and green space, and the connection with nature was driven out of these growing cities. While the countryside provided proximity to nature and an abundance of ecosystem services, it also suffered from a lack of jobs and economic prospects. As a way to address these issues, and to create healthier towns and cities in which to live, in the late-nineteenth century the concept of Garden Cities was devised by Sir Ebenezer Howard (1850–1928). This model of urban planning was intended to solve the issue of rural flight and the uncontrolled growth of urban areas through the creation of a series of small cities that would combine the benefits of both environments. Howard is best known for his book '*To-morrow: A Peaceful Path to Real Reform*' (1898) which was later republished as '*Garden Cities of To-morrow*' in 1902, which set out his ideas of social and urban reform.

The Garden Cities were intended to avoid the pitfalls of the newly industrialised cities, which included low wages and poverty, overcrowding, inadequate drainage, dirty streets, poorly ventilated houses, smog and air pollution, and the respiratory and infectious diseases that accompany these, not to mention the mental health implications of a lack of interaction with nature. The Garden Cities offered a vision of towns free of slums, with all the benefits and diversions of up-and-coming urban areas, such as jobs, good wages and social diversions and amusements, while at the same time providing the positive effects of the country, such as natural beauty, fresh air and lower population density, and hence lower land costs. This idea was illustrated by the now famous Three Magnets diagram (figure 6.1) which considered the issue of where people will go. Each magnet represented a specific environment, Town, Country or Town-Country. The Town and Country magnets list the positives and negatives of each environment, while the final Town-Country magnet combines the advantages of both. This premise was the foundation of the Garden Cities concept.

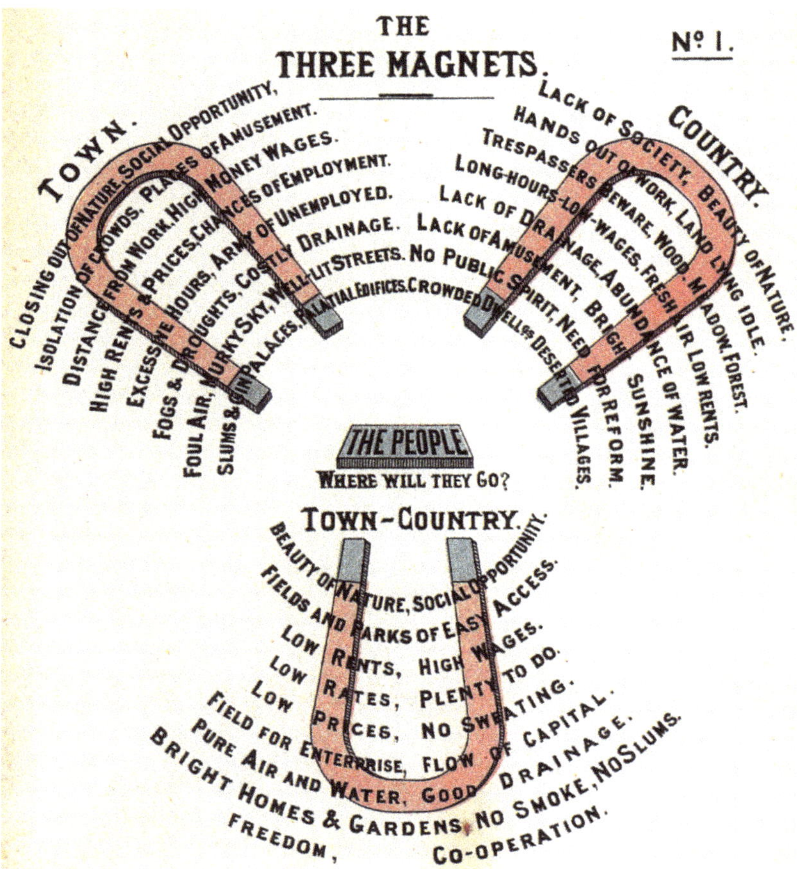

Figure 6.1. The Three Magnets diagram originally published by Ebenezer Howard in '*To-morrow: A Peaceful Path to Real Reform*' (1898).

Howard proposed that society should be reorganised according to networks of smaller garden cities that would break the consumerist nature of larger cities. The *Garden Cities of To-morrow* book proposed the creation of new suburban towns of limited size, planned in advance to prevent sprawl and surrounded by a permanent greenbelt of agricultural land.

These Garden Cities were used as the model for many suburbs where the aim was to blend city and nature. The towns were intended to be largely independent and managed by citizens who had an economic interest in the success of the town. The land on which the towns were built were to be owned by a group of trustees and the leased to the citizens.

It is often assumed that the because the diagrams in Howards books show circular cities (see figure 6.2) that the intention was for the cities themselves to be circular. However, the diagrams presented in '*To-morrow: A Peaceful Path to Real Reform*' and '*Garden Cities of To-morrow*' were not intended to be the basis for physical plans of cities. Indeed, Howard notes in these books that each city should be organised as per the needs of the people and the local environment, and should not be circular.

A key part of the Garden City concept was the distribution of land use. Industry was to be located around the outside, while amenities such as schools and parks were to be located centrally, as shown in figure 6.3. The key concept was to encourage a connection with nature that was missing in the cities of the time, the actual layout did not have to follow this blueprint, but rather just its ideals, with the aim of promoting health and wellbeing amongst its citizens.

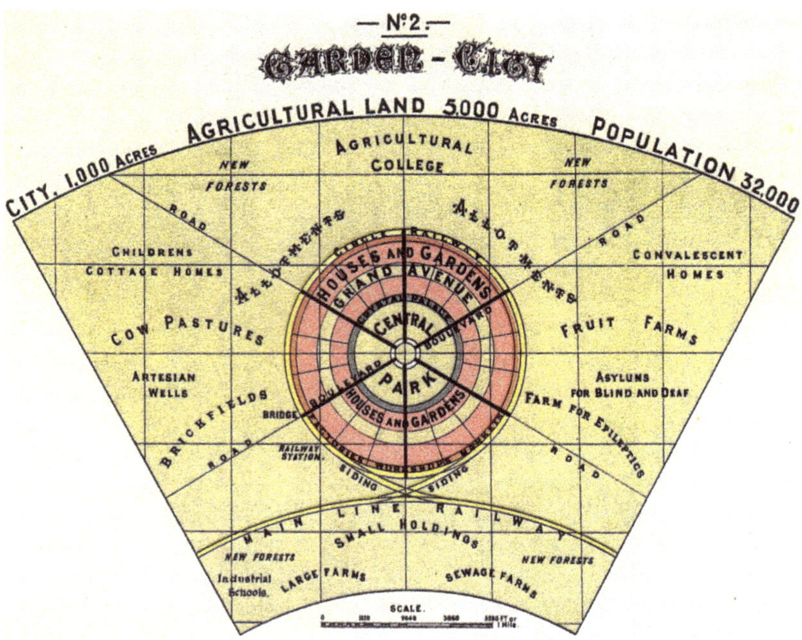

Figure 6.2. Schematic of a Garden City, showing the suggested scales and land uses, published by Ebenezer Howard in '*To-morrow A Peaceful Path to Real Reform*' (1898).

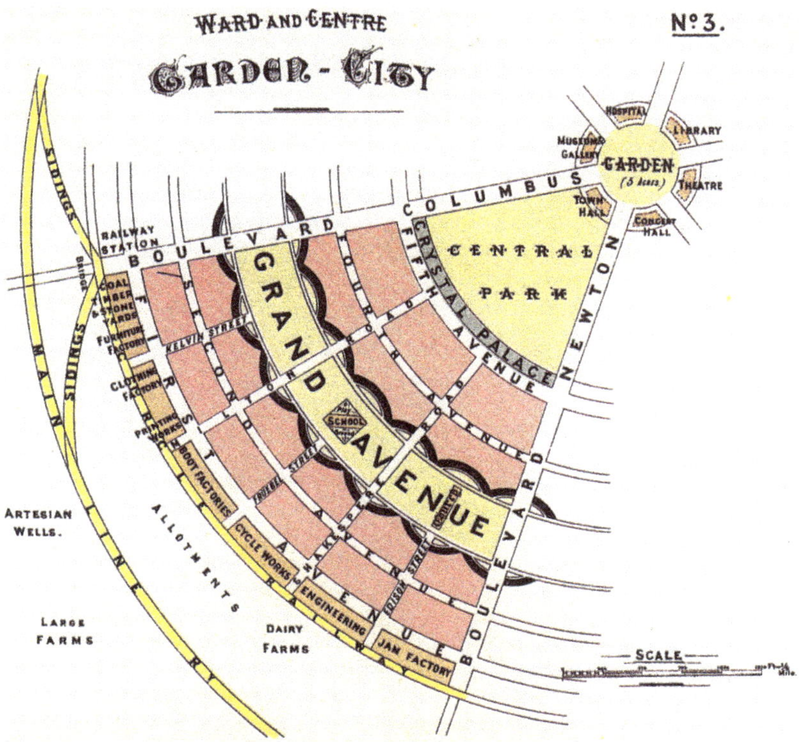

Figure 6.3. A sector of a Garden City showing the distribution of land uses, published by Ebenezer Howard in '*To-morrow A Peaceful Path to Real Reform*' (1898).

Due to the limited size of the intended Garden Cities several specialised cities would be located close together within easy commuting distance to create a network. This network was Howard's visualisation for replacing the large, polluted cities of the late-nineteenth century. The interspacing of the orbital towns with greenspace, together with the efficient provision of transport infrastructure between the towns and the central city, was key to the concept (see figure 6.4).

Howard's ideas attracted sufficient attention and funding to begin Letchworth Garden City in 1903, a suburban garden city located ~37 miles (~60 km) north of London. Letchworth was an existing parish that appears in the Doomsday Book and had remained a small rural village until the land and its environs was purchased by Howard's company 'First Garden City Limited'. Letchworth Garden City was designed to house a population of between 30 000 and 35 000 and in the last census had a population of 33 249. The city itself was laid out as per the instructions in Howard's book, there was a central town along with agricultural, commercial and industrial areas, distributed civic centres and open spaces. This division of land is now referred to as zoning and is now a commonplace concept in urban planning.

Howard constructed Letchworth Garden City as an example of what could be achieved, while some criticise Letchworth for being too spacious and inefficient in land use, it could be argued that this is what makes Letchworth a pleasant healthy

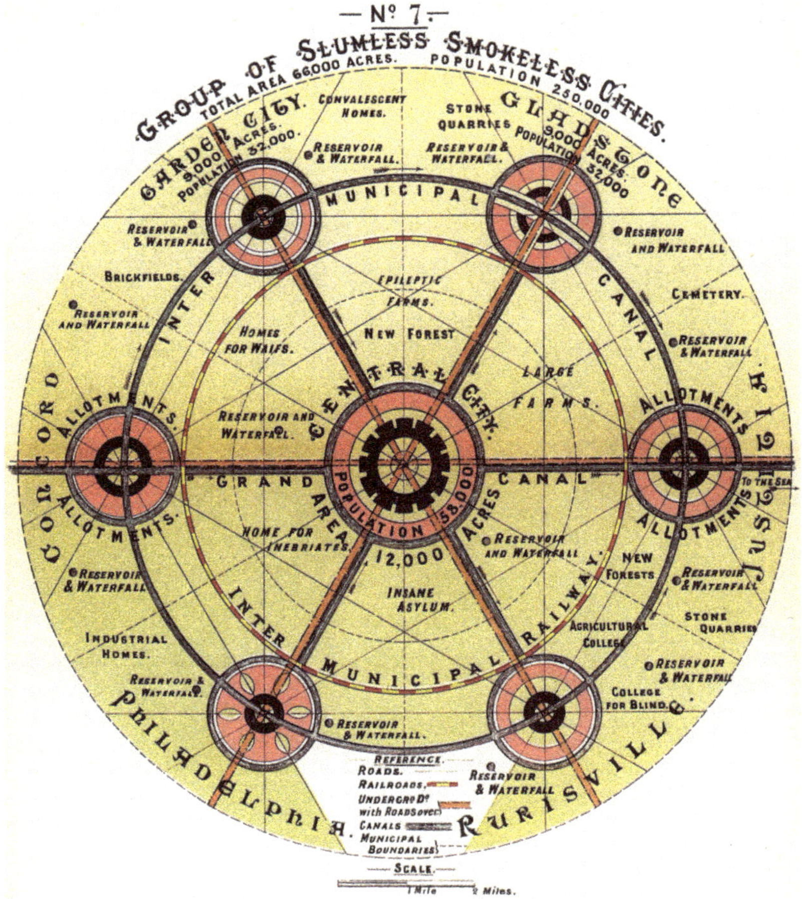

Figure 6.4. Networks of several small garden towns surrounding a central city, interspaced with green space would replace a single large city. Published by Ebenezer Howard in '*To-morrow A Peaceful Path to Real Reform*' (1898).

place to live. Preparations for a second garden city, Welwyn Garden City, located some 20 miles (32 km) north of London, began in 1919 following World War I. After 10-years of existence, Welwyn Garden City was home to ~10 000 people, again with distinct residential, commercial and industrial zones, today Welwyn Garden City is home to 48 380 people. The town itself is laid out along tree-lined boulevards radiating from a town centre and every road has a wide grass verge. By 1930, the health of the Welwyn Garden City residents was considerably better than those who lived in London, with consistently lower death and infant mortality rates. The reasons for this are generally understood to be as a direct result of the Garden City design principles.

The creation of Letchworth and Welwyn Garden Cities and their associated benefits were influential in the British government's proposals for the development of new towns and rebuilding after World War II. This resulted in over 30 new communities, culminating in the largest Garden City to date at Milton Keynes ~50 miles (~80 km) north-west of London, which today houses 264 349 people. It is clear that the Garden City movement resulted in some special urban areas during periods of population growth and rebuilding in Britain's history. While the Garden Cities were intended to solve the socioeconomic problems of the late-nineteenth century, it can be argued that these problems still exist now in the rapidly urbanising Global South. The benefits of more dispersed urban areas with significant areas of open space will be of benefit when we consider the impacts of climate change.

6.3 Urban geometry effects on comfort and energy use

As we discussed in the previous chapter, urbanisation whether that is urban growth or densification, not only transforms the spatial and socioeconomic distribution of the local population, whether that is the urban or the surrounding rural population but it also changes the energy budget of the built environment. Densification of the built environment can be typified by narrower urban canyons (relative to their height) with greater use of impervious materials, less green and blue space, and increased pollution. This has the knock-on effect of increased sensible heat storage, reduced sensible to latent heat conversion, the trapping of shortwave radiation within the street canyon and reduced sky view factors for the emission of longwave radiation (heat). All these factors give rise to a distinct urban heat island, which, through modification of the energy budget, will have an effect not only on the energy performance of buildings but also on the thermal comfort of pedestrians in the urban area.

Outdoor thermal comfort is recognised as a key performance indicator of urban environmental assessment because it has been shown to influence the travel choices and activities that pedestrians undertake. There are, however, few studies that consider outdoor thermal comfort, most deal with thermal comfort within buildings rather than outside them. This is mostly due to the large spatial and temporal fluctuations in temperature, air speed, radiant temperature, etc that can occur across an urban area. These environmental variables are the determinants of pedestrian comfort, and will determine how people react both physically and physiologically in order to maintain their own thermal equilibrium and comfort level. Given the complex interdependent relationships between the size, density, geometry of a city and both the urban microclimate and building energy consumption, we can state that the built environment has a role in addressing the environmental issues posed by climatic change, rather than just being a cause of it.

The relationship between climate and architecture is typically one way, architects and urban planners will consider the background climate and design buildings and urban areas appropriately to respond to the environmental conditions. But does this need to be the case? There are a few studies that have considered how the built geometry and urban form affect the different pertinent environmental variables such as wind ingress, solar exposure and daylighting. From these, we can also estimate levels of thermal discomfort and anticipate changes to the urban heat island effect and the local microclimate. The most obvious example of this effect is the use of courtyards in hot-arid climates to provide shade for living areas and trap cooler air inside the building. Such design decisions, however, cannot be used with impunity. A study of the courtyard forms in in the oceanic climate of Stuttgart, Germany demonstrated up to a 20% increase in heating energy demand. On a wider scale the work of Oke [1] studied the use of generic urban forms and their morphological parameters linking them to local environmental conditions, and proposed a simplified classification of urban climate zones based upon the built configuration, terrain roughness, the aspect ratio and the level of ground imperviousness. This led to the creation of 17 local climate zones (LCZs), which can be used to delineate regions of uniform-built types, land cover, materials and human activity that can extend up to several kilometres. The homogeneity of these LCZs is unlikely to be found in the natural world, but one would expect the 17 distinct patterns to be familiar to residents of most cities.

Studies of the external temperatures, UHI intensity and heating and cooling demand for five different urban geometries in the cities of Rome and Barcelona conclude that the compact (horizontally dense) urban geometry with over 50% site coverage leads to lower annual energy consumption compared to rural areas, but they also exhibit higher UHI intensities in the wintertime. The use of parameters associated with the scale and geometry of buildings is critical to be being able to classify various urban studies. These include building heights (H), street widths (W) and the associated aspect ratio (H/W), together with the sky view factor (SVF) and orientation determine the performance of buildings and the urban street canyons in terms of temperatures, comfort levels and seasonal energy loads. At a neighbourhood level, these parameters can be amalgamated into concepts such as floor area ratio (FAR) and building coverage ratio (BCR).

The integration of the various tools needed to model both building performance and environmental parameters at an urban scale is still a challenging prospect due to the considerable amount of computational power required. As such, urban geometry studies over the last two decades can typically be characterised into either studies investigating the impact of urban geometry on energy consumption, solar radiation and daylighting, or studies concerned with urban geometry, the UHI and the microclimate. Very few studies consider both aspects!

The work of Ibrahim *et al* [2] studied the effect of various urban geometries and densities on both building energy consumption and urban thermal comfort in the hot-arid climate of Cairo, Egypt. The authors used abstracted building typologies and standardised design parameters to reduce the complexity of the parametric simulations. Hypothetical urban blocks comprising of arrays of residential buildings, 21 m × 21 m each, were modelled within an identical surrounding context. This gave rise to buildings with a floor area of 441 m^2 comprising of four apartments, which are comparable with existing building typologies in the region. Three different basic typologies were considered: courtyard, linear and scattered, as shown in figure 6.5.

The scattered building layouts were simulated with street canyon widths ranging between 6 and 18 m, while the distance between individual buildings was set as half the street canyon width. As the distance increases beyond 18 m in each direction,

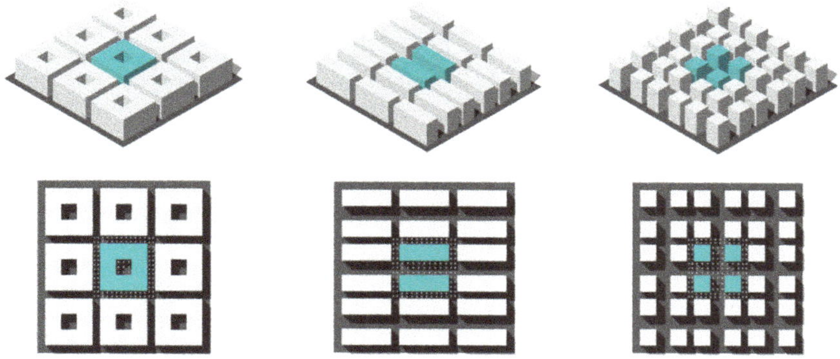

Figure 6.5. Three building archetypes studied by Ibrahim *et al* [2] together with their surrounding urban context (top) and a plan view showing sensor locations for external comfort (bottom). Reproduced from [2]. CC BY 4.0.

new forms of linear or courtyards layouts were created (i.e. at 21 m separation a new building was added in the gap to form a linear arrangement, the same gap in both directions creates a courtyard block of eight buildings).

Ibrahim *et al* compared these different building typologies and the corresponding urban morphological parameters using environmental, energy and comfort metrics. Energy consumption was accounted for using the energy use intensity (EUI), which is the final energy consumption per unit area. In this case, cooling, heating, lighting and electric plug loads were considered. Outdoor thermal comfort was estimated using the universal thermal climate index (UTCI) metric. UTCI is widely used for outdoor studies and presents a categorisation of thermal sensation ranging from severe cold to extreme heat conditions. UTCI is a complex metric, comprising of a variety of climates variables and human factors such as clothing and activity level, similar to how PMV is estimated for internal environments. UTCI was recorded for a number of sensor points located in the street canyons at a height of 1.1 m to represent the centre of mass of a typical human. UTCI was recorded between 5 am and 7 pm, and averaged across all the canyons and then again over the month of July (the hottest month in Cairo) this was then compared to monthly EUI averages.

The energy and comfort performance of each building typology was calculated for a range of building separations and orientations. Figure 6.6 shows a frequency distribution for each of the performance metrics considered for each of the building typologies (i.e. scattered, linear and courtyard). We can see that the courtyard and scattered forms have more instances of lower average UTCI. The linear forms, on the other hand, have more instances in the mid-range average UTCI values than the other two typologies. It is interesting that the scattered and linear forms also have the highest average UTCI values. This indicates that the variability of the scattered form in terms of external thermal comfort is considerable, while the courtyard forms are more consistent in maintaining lower average UTCI values. This is confirmed in panel (b) of figure 6.6 with the courtyard form having a clearly lower distribution of max UTCI than the scattered or linear forms. However, the lowest values of max

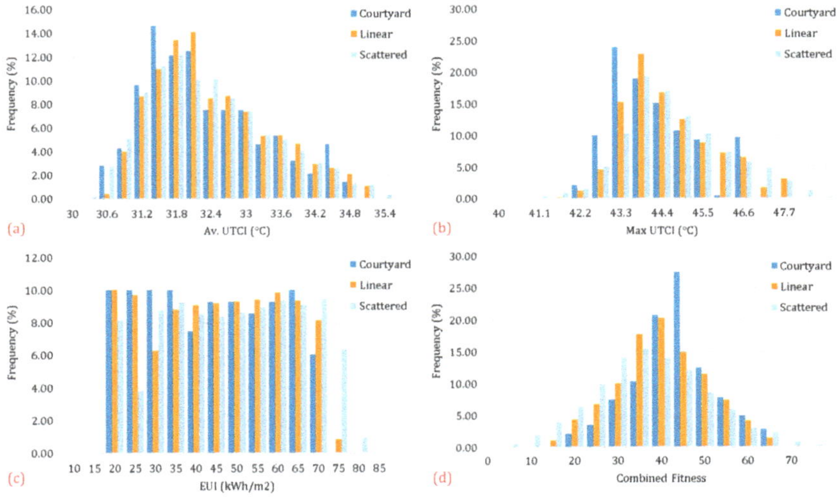

Figure 6.6. Frequency distribution for each of the building typologies for (a) average UTCI, (b) max UTCI, (c) EUI and (d) combined fitness. Reproduced from [2]. CC BY 4.0.

UTCI belong to the scattered typology, indicating that compact high-density scattered forms can perform as well if not better than courtyards due to the shading they cast in both street directions.

Panel (c) of figure 6.6 shows the EUI performance of the different forms in kWh m^{-2}. We can see that the courtyard and linear typologies have the lowest EUI, most likely due to greater mutual shading effects, resulting in lower internal temperatures and cooling loads, and resultantly a lower EUI value. Consequently, the scattered forms dominate the higher EUI frequencies due to their greater exposure to solar radiation. Panel (d) shows the combined fitness of each building form based upon the comfort and EUI metrics. The scattered forms are skewed towards a lower fitness value and the mean of the distribution is two units below the composite average of 40 across the three forms. While the linear forms demonstrate the greatest values of combined fitness, the average of this distribution is only one unit above the composite average, compared to four units for the courtyard forms. This indicates that in terms of EUI reduction and lower UTCI values, courtyard forms consistently perform better than the other typologies in the hot-arid climate of Cairo.

During the month of July considered by Ibrahim *et al* [2] the solar altitude in Cairo reaches as high as 83°, meaning that north–south orientated street canyons are shaded in the morning and late afternoon but are fully exposed to high levels of solar radiation across the middle of the day. The high solar altitude also means that many east–west orientated street canyons are exposed to direct solar radiation over the entire course of the day, with only minimal shading around noon. This makes the cardinal compass directions less desirable for street canyon orientation, instead ordinal orientations are preferable for lower solar ingress as they provide shading in both directions. This translates to lower average UTCI values, as shown in panel (a) of figure 6.7. It should be noted that NE/SW orientation is equivalent to NW/SE due

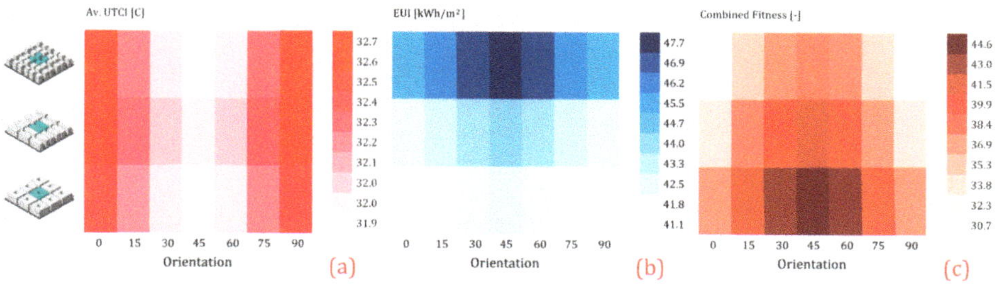

Figure 6.7. Comparison of the performance of each of the building typologies versus street canyon orientation (north/south = 0°, east/west = 90°). Reproduced from [2]. CC BY 4.0.

to the symmetry of the simulated urban area. We can see that the courtyard forms demonstrate slightly lower average UTCI values than the scattered forms, and both are significantly lower than the linear form across all orientations. This, however, is more complex than it appears at first glance, east–west facing buildings (i.e. north/south canyons) will tend to be more exposed to solar radiation than north or south facing walls. Again, courtyards and linear typologies have greater mutual or self-shading. This results in the orientation performance of the different typologies in terms of EUI being very different to the UTCI results, as shown in panel (b) of figure 6.7. This means that in terms of a combined fitness (panel (c) of figure 6.7), the courtyard forms orientated close to the ordinal directions have clearly better performance than other orientations and typologies. The scattered forms, regardless of orientation, perform worse when considering both comfort and energy performance. The results of this study indicate that orientation alone is responsible for almost a 15% change in average UTCI and a 10% difference in energy consumption. It has previously been shown that orientation and surface to volume ratio of a building have the greatest influence on annual energy loads in a Mediterranean climate. This indicates that the same also holds true for the hot-arid climate of Cairo.

Ibrahim *et al* [2] also investigated the effects of building height on the performance of the different building forms. This alters the H/W ratio in the same way that altering the street width does, but it also increases the floor area ratio (FAR) without altering the BCR. Building heights were varied between 9 and 36 m in increments of 3 m indicating an additional storey. The changes to the buildings' heights was shown to have the strongest correlation to the thermal and energy performance (see figure 6.8). The effects of changing the height of the buildings would appear to be more significant than changing the typology. As building height increases, we can see that the average UTCI decreases by nearly 2.5 °C, this is due to increased shading, reducing mean radiant temperatures, which improves the level of thermal comfort in this hot-arid climate. The relationship between average UTCI and building height (and hence FAR) appears to be asymptotic, indicating a maximum possible benefit, which in this case occurs at 33 m building height, after which streets become fully shaded. Interestingly, the difference between the three typologies is minimal in terms of average UTCI.

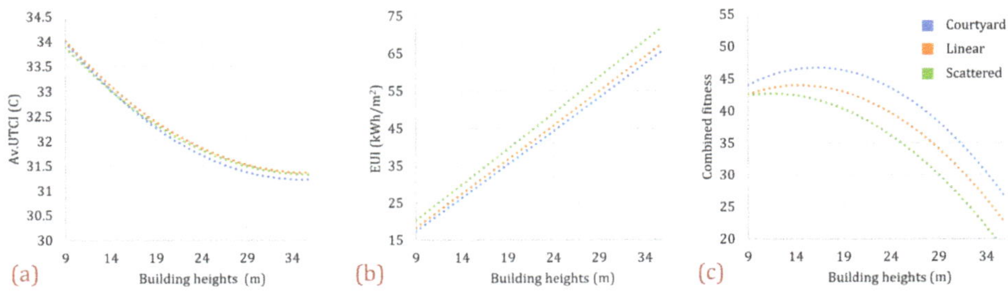

Figure 6.8. Trendlines of building performance versus height for the different building typologies. Reproduced from [2]. CC BY 4.0.

Panel (b) of figure 6.8 shows that the influence of building height on the EUI is profound, increasing by as much as 50 $\mathrm{kWh\,m^{-2}}$ with increasing building height. This impact is much more significant than the effect on average UTCI and so the variation between typologies, although this is still small by comparison. This large linear relationship between EUI and height alters the shape of the combined fitness curve, which indicates a distinct sweet spot (see panel (c) of figure 6.8). The combined fitness of EUI and average UTCI with building height indicates that after a height of 24 m, increased shading is no longer able to overcome the effects of increasing EUI and the fitness declines sharply. Similar trends have been reported in hot-humid climates, where it was concluded that the albedo, *H*/*W* ratio and the distance between buildings can command up to an 80% variation in solar exposure, leading to large variations in comfort and energy use.

The conclusions of Ibrahim *et al*'s [2] study shows that there is a statistically significant correlation between the design parameters and the performance indicators, average UTCI and EUI. These metrics display large variations attributed to changes in the urban density parameters, FAR and BCR. Interestingly, the altered design parameters showed impacts on the different typologies that seem similar in nature abut are disparate in magnitude. For instance, changing street width and changing building heights independently both alter the *H*/*W* ratio of the street canyon but the impact on average UTCI and EUI are not the same. Similarly, combinations that yield the same BCR and FAR are not equivalent. Compact and dense scattered building forms performed as well or even slightly better than courtyard typologies, both outperforming linear configurations when considering outdoor thermal comfort. Conversely, courtyards and linear building forms performed better than scattered forms in terms of energy performance due to the ability of courtyard and linear forms to self and mutually shade building facades. Another aspect of this is undoubtedly that the scattered forms have a larger surface area to volume ratio, increasing heat exchange with the external environment. Orientation was shown to have a considerable impact upon the overall performance. From a combined comfort and energy point of view, compact (BCR$\geqslant$60%) and medium-density (4$\geqslant$FAR>2) courtyards orientated at 45° showed the best combined performance.

The results presented by Ibrahim *et al* [2] indicate that there is great potential for urban planners and designers to create healthier and more resource efficient urban communities, which are ultimately more resilient in the face of rising global temperatures. Ibrahim *et al*'s [2] study focused upon the hot-arid climate of Cairo, Egypt; however, the results would be largely applicable to any hot-arid climate at a similar latitude. Moreover, the methodology used in this study could easily be applied to other climatic contexts. It is expected that the optimum geometric form and density will vary with location, due to solar height and background climate. For instance, it might be expected that scattered forms perform best in hot-humid climates where more ventilation is required. Greater diffuse light from the sky dome due to increased humidity levels may well reduce the effects of mutual shading, with isolated standalone buildings with large roof overhangs to provide self-shading being optimal. The exact geometric dimensions of the optimal urban area would require further research but the potential to compare many different urban geometries within a computer program and find the best solution based upon energy and comfort requirements for any location is so far an untapped resource.

6.4 Green and blue infrastructure

Green and blue infrastructure or green and blue space refer to vegetated surfaces or bodies of water (static or dynamic) within an urban area. Green space can take many forms in an urban area, whether grassy parks or urban forests, tree-lined avenues or grass verges, private gardens or the fringes of transport corridors. These green and blue spaces provide many varying ecosystem services to the urban area. These services include reduced surface runoff, flood relief, sustainable drainage, increased biodiversity as well as enhancing the general aesthetic and improving the health and wellbeing of the local populous. In this chapter however, we are primarily concerned with the ability of green and blue spaces to modify the local microclimate. A recent study of Glasgow (Scotland) suggested that an increase in green space around the city of ~20% could eliminate between 30% and 50% of the city's UHI. The introduction of strategically planned and interconnected networks of green and blue spaces, offering ecological, social, economic and climate resilience benefits are referred to in city planning as green or blue infrastructure.

The microclimate benefits of green and blue space arise from the evaporation and transpiration of water (collectively evapotranspiration), transferring energy from the surface to the atmosphere, thereby linking the urban energy balance with the hydrological cycle. It is estimated that the process of evapotranspiration consumes ~22% of the total solar radiation impinging the top of the atmosphere. The reduction in evapotranspiration associated with urban areas alters the energy balance and heat that would otherwise have been consumed as latent heat, heats the urban surface and contributes to the formation of the UHI.

Mitigation of the UHI can be achieved through the use of green and blue infrastructure. The addition of vegetation and bodies of water to an urban area, increases evapotranspiration, thereby replacing sensible heating Q_h (increasing temperatures) with latent heating Q_e (phase change of water). This reduces the *Bowen ratio*, which is the ratio of sensible to latent heat flux ($B = Q_h / Q_e < 1$), to produce evaporative cooling.

6.5 Thermal effects of green space

Green space affects the urban thermal energy balance in both directly and indirectly. It directly influences the climate by reducing surface and local air temperatures, which in turn affects wider climate air temperatures. It also indirectly modifies it by reducing heat transfer into occupied spaces, and thereby reducing cooling loads and any resulting anthropogenic heat emissions back into the UCL. The most discussed of such vegetation cooling processes is transpiration, where water transported through the plant is evaporated at the aerial parts by absorbing energy from solar radiation that increases latent rather than sensible heat to keep the foliage cool, and thereby cooling the surrounding atmosphere. For most vegetated surfaces, 99% of the water and over half the solar radiation absorbed is typically used for transpiration.

The species of plant considered is significant if the aim is to mitigate the UHI. While C3 photosynthesis plants typically found in cool and wet climates transpire significant volumes of water, C4 photosynthesis plants that are adapted to warmer climates transpire much smaller volumes of water. Drought-tolerant plants typically found in arid climate use crassulacean acid metabolism photosynthesis and minimise water loss by keeping their stomata closed during the day, only opening at night. This effectively minimises daytime cooling potential due to their negligible transpiration rates. In the majority of plant types (C3 and C4), leaf stomata are typically closed in the absence of solar radiation (at night). Latent cooling from transpiration is therefore primarily a daytime energetic process. The rates of transpiration achieved during the day depend on the characteristics of the plant in question: crown area, leaf area index (LAI) defined as the single surface leaf area per unit of ground area, height of the leaves, stomatal resistance, the hydraulic resistance of the shoot and roots and the local soil conditions described by dryness, compaction and hydraulic conductivity, all have an impact on the transpiration rates achieved. The local climatic conditions also influence transpiration rates because plants constrict or close stomata in order to control heat stress and water loss. A reduction in the cooling effectiveness of plants subsequent to protracted droughts or high temperature conditions is to be expected. This presents a problem when considering using green space to reduce the UHI and provide protection for urban areas against the higher temperatures expected as a result of climate change. The increase in temperatures and the shift in rainfall patterns from summer to winter will reduce the effectiveness of plants to provide evapotranspirative cooling. Furthermore, the use of drought-tolerant plants from more arid climates, which is commonly cited as a suitable response, will not solve this issue because these have a lower transpiration rate in the first place. At night and during periods of high temperature, or when there is insufficient water, plants respire like other lifeforms rather than photosynthesise, emitting CO_2 and generating heat. Such effects are taken into account in global climate models as a feedback process, but this is given little consideration in urban climate discourse, which often opts for a blanket 'more green is better' approach.

As discussed in chapter 5, shading from vegetation keeps the atmosphere cooler through intercepting solar radiation and by reflecting energy due to its lower albedo or absorbing energy to drive photosynthetic processes, and thereby limiting shortwave absorption by urban surfaces (figure 6.9). This reduces urban surface temperatures and the eventual longwave radiative purging into the canopy-layer atmosphere. The reflection of radiation reduces shortwave absorption, with grass credited to reflect ~20%–25% and trees 15% of the incoming energy back to the atmosphere. A significant proportion of the absorbed shortwave energy is utilised by phyto-active chemicals in vegetation for biological photosynthesis, while the residual is held in storage. The effectiveness of vegetative shading on a surface is determined by the leaf size, crown area and LAI of the vegetation canopy. As you might expect, trees and to lesser extent shrubs are more efficient at shading than species of grasses. A tree canopy can therefore create a localised microclimate beneath it, so-called trunk-space cool spots.

The canopies of vegetation alter the surface roughness, which in turn modifies wind flows and the convective heat exchange. The type of vegetation and the canopy density are particularly significant here, with grasses and low-lying shrubs creating a barrier of stagnant air near the ground and dense clusters of trees or forests impede wind flow to trap warmer insulated air beneath the canopy. However, dispersed groves of varied vegetation provide canopy heterogeneity, enhancing surface roughness, generating turbulence and increasing convective heat loss. With lone

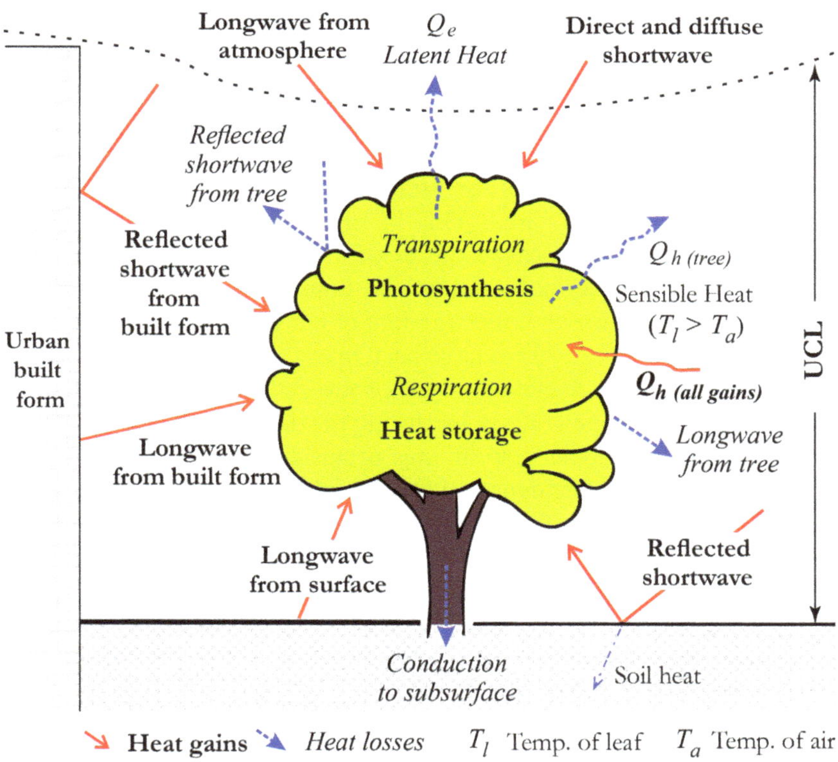

Figure 6.9. Energy exchanges between green space and the urban environment.

isolated trees, the convective heat loss tends to be greater because they protrude through the UCL into the UBL, which increases exposure to drier air flowing from non-vegetated (urban) areas. This increases the water vapour gradient and increases evapotranspiration through leaf stomata.

In addition to these processes of evapotranspiration, solar shading, pollution filtering and modification of wind flows, vegetation also indirectly cools the local climate. PM pollution filtering is achieved by dry deposition, a process where the pollutant molecules or particles impact upon and stick to vegetation surfaces such as canopy leaves. Removing PM from the atmosphere reduces atmospheric scattering and absorption of shortwave solar radiation and longwave infrared radiation emitted by surfaces. This alters the radiation balance and the rates of atmospheric warming or cooling, typically a clearer atmosphere will allow surfaces to not only warm faster but also cool faster. Vegetation canopies also reduce surface runoff by the interception of rainfall. The typical softer landscaping underneath vegetation also aids in the reduction of runoff rates and encourages greater absorption, this increases soil moisture allowing for prolonged evaporative cooling after the rainfall event.

The cooling effect experienced depends upon the environmental conditions of the vegetated area. Background soil moisture content is particularly significant with precipitation or irrigation providing greater soil water for transpiration, while high atmospheric humidity reduces the water vapour gradient and suppresses transpiration. Increased moisture availability also tends to determine the types of vegetation that result, with greater moisture availability resulting in denser growth increasing surface roughness compared to drier climates. Wind flow is advantageous in high humidity conditions because it advects away accumulated saturated air, with higher wind velocities reducing the leaf boundary layer to enhance the water vapour gradient, increasing latent heat flux. These environmental variables of moisture content and wind flow along with the ambient temperature influence the types of vegetation native for a given area. This in turn determines the availability and effectiveness of the cooling processes discussed above.

In order to assess the thermal comfort and human health benefits of increasing green space, we need to have an understanding of the spatial extent of the impacts of any urban greening proposal. There have been many studies of the impacts of urban parks, but little consensus. A meta-analysis of several studies in the UK found that urban parks were on average 1 °C cooler during the day, but that the extent to which this cooling permeated in the surrounding urban context varied considerably. A 1960s study of Kensington Gardens and Hyde Park in London found a 3 °C cooling effect, which extended up to 200 m from the park boundaries. A more recent 2014 study of Kensington Gardens showed a mean summer temperature reduction of 1.1 °C with a maximum reduction of 4 °C observed on certain nights. The distribution of nocturnal cooling range was observed to be between 20 and 440 m, with 83% of the cooling potential still evident 63 m (the median of the range distribution) from the boundaries. The cooling influence of green space observed varies significantly both spatially and temporally. If vegetation is to be utilised for the cooling of urban areas, we need to understand the factors that determine the distribution of cooling (both horizontally and vertically) into the surrounding urban context.

Although numerous studies of urban parks have demonstrated the horizontal distribution of their cooling effects, there is little quantitative evidence to allow us to determine how isolated parks will affect the overall climate of a city. This issue is highlighted in studies of London. In contrast to most large cities, London is relatively green, with ~47% of its total area considered green, 33% vegetated green space and 14% as planted domestic gardens. However, a comparison of surface temperatures and the average atmospheric heat island for relatively warm summers (observations and simulation data output from the LUCID project, led by University College London) reveals complex cooling distribution patterns (figure 6.10). Accounting for prevailing south-westerly winds, several areas of interest can be identified. Although cooling from Kensington Gardens and Hyde Park is apparent at the surface level, at the higher atmospheric level its influence is not apparent. Comparison of these images shows that combined green spaces of Richmond Park and Wimbledon Common and to a lesser extent Hampstead Heath are the only green spaces to have an influence in the higher atmosphere.

If we overlay the atmospheric (UBL) heat island data on top of a map of London showing green and blue spaces (figure 6.11), we can see that several large green spaces such as Hyde Park, Kensington Gardens and Lee Valley Park appear to have no influence on temperatures in the higher atmosphere. There is also no visible influence from the river Thames or from other large blue spaces; this, however, will be discussed later in the chapter.

From figure 6.11, we can see that the cooling influence of green space depends on more than just area. If pure size was the only variable in the cooling influence, then Lee Valley Regional Park would produce the greatest effect; however, its UBL cooling influence is not visible. It can be hypothesised that the magnitude and

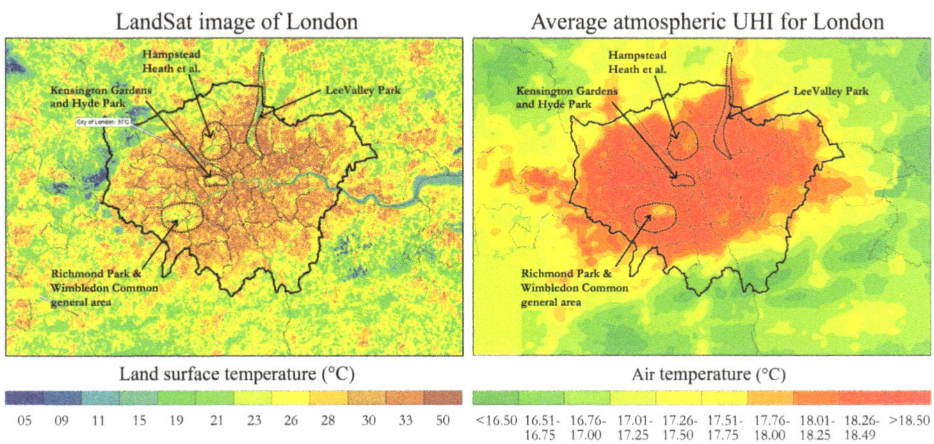

Figure 6.10. Surface temperatures from LandSat data (left-hand panel) and UHI simulation data from the LUCID project (right-hand panel)[1].

[1] Details on the creation of figures 6.10 and 6.11 can be found in [3].

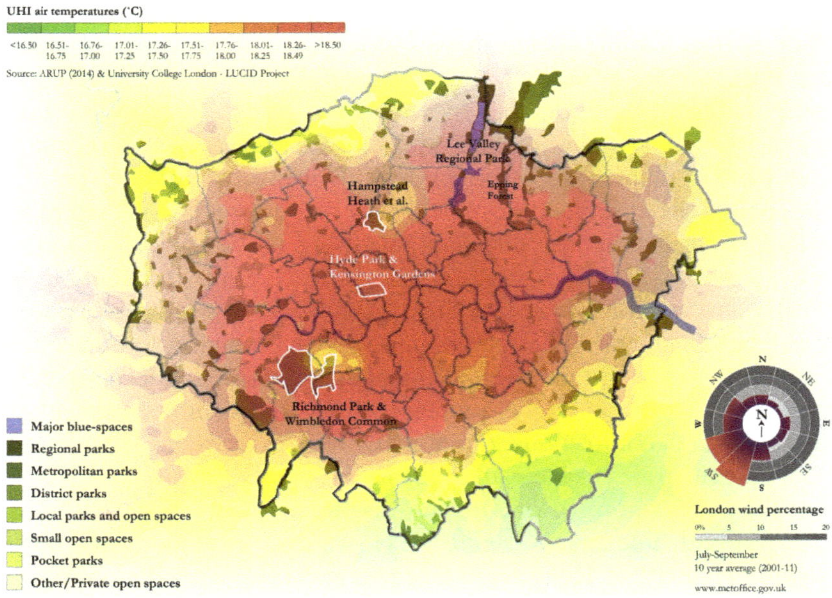

Figure 6.11. UBL UHI overlaid over a map of London's green and blue spaces. Reprinted from [3], Copyright (2017), with permission from Elsevier.

Table 6.1. Major green spaces in London and their approximate dimensions

Green space	~Area (km^2)	~East–West span (km)	~North–South span (km)
Hyde Park + Kensington Gardens	2.5	2.5	1.0
Hampstead Heath	3.2	1.7	1.8
Richmond Park	9.6	4.0	4.5
Epping Forest	24.8	2.7 (max)	8.8 (linear)
Lee Valley Regional Park	40.5	1.4 (max)	42.0 (linear)

geometrical distribution of the green space has a significant bearing on the citywide (UBL) cooling expected. Table 6.1 shows the approximate dimensions of London's major green spaces. It is possible that the relatively linear geometry and limited fetch (for south-westerly winds) of the Lee Valley Regional Park (~1 km wide, compared with 5 × 7.5 km for Richmond Park and Wimbledon Common) impedes the development of the strong temperature and humidity gradients necessary to produce significant vertical transport and provide citywide cooling. Another possible contributing factor could be that a significant proportion (~22%) of this park is blue space, taken up by reservoirs (the significance of this is discussed later).

This hypothesis in relation to Hampstead Heath, which is relatively small together with its surrounding context, suggests that clustering of green spaces with suitable surface roughness is able to produce the vertical transport necessary to effect citywide cooling. There is currently a significant gap in the academic literature regarding monitored vertical distribution data that prevents testing of the relationship between geometric parameters and the vertical transport of cooling within the UBL. The majority of studies present data in relation to canopy-layer cooling and almost entirely focus on the horizontal distribution of cooling. However, even in the absence of solid empirical evidence, it seems certain that the size, shape (determining fetch for different wind directions) and the diversity of green spaces (determining surface roughness) influences the ability of a park to affect the UBL UHI and provide mesoscale (citywide) cooling.

6.6 Green space implications for city planning

Many models are used by city planners to describe urban growth that describe contrasting land-use morphologies and distributions. Of the many such models, the two theoretical extremes are the *compaction* and *dispersal* models (figure 6.12). Both feature widely in city planning discourse and there is a general preference for the compaction model because it is believed to offer a more sustainable framework for growth. This preference for urban containment has its origins in the desire to safeguard the countryside from urban encroachment. The aim of the compaction growth model is to promote the efficient use of land and infrastructure. Higher density is generally considered to reduce resource consumption and lower economic and environmental costs (CO_2 emissions). The use of compaction policies is evident in modern England, where ~80% of the population occupies only ~20% of the land, with over 70% of new developments taking place on brownfield sites at relatively high densities.

In contrast, the dispersal growth model is promoted by free-market advocates who argue that if land-supply (greenbelt) restrictions are lifted, then land will become more affordable and hence more socially accessible. Evolving socioeconomic patterns seem to favour the dispersal model, with many European cities

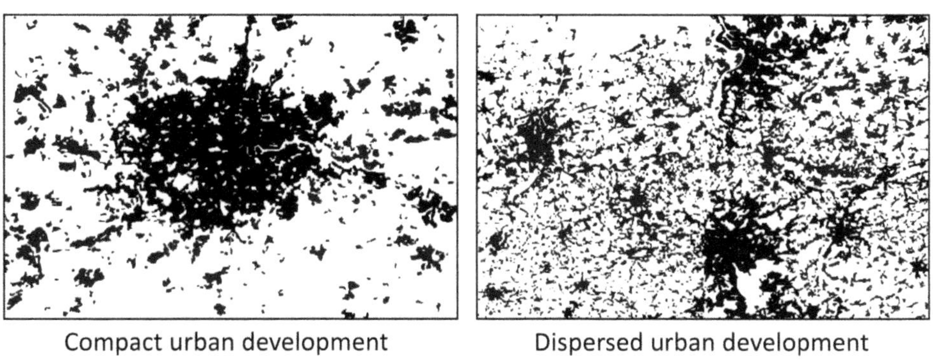

Figure 6.12. Illustration of the compaction and dispersal urban growth models.

moving from single urban centres to more polycentric cities. However, the dispersal approach is treated with scepticism by policymakers because allowing market forces to manage land-supply and demand may well lead to urban sprawl and low-density developments, resulting in a high dependency on resource use. Furthermore, dispersal is associated with increased energy consumption, adversely affecting greenery and biodiversity, requiring extensive infrastructure networks.

Dispersed urban development is often criticised for increased land use compared to compaction or densification strategies, with urban growth likely to be on greenfield land leading to the loss of peripheral green space. A US study showed that the rate of rural green space loss in the most dispersing urban areas was more than twice that in the most compact urban areas, with association made between the frequency of extreme heat events and the loss of regional vegetation. Another study of Frankfurt (Germany) demonstrates the importance of safeguarding peripheral green space, showing a beneficial cooling effect of between 3 °C and 3.5 °C. The Frankfurt study suggests this cooling influence occurs due to the formation of a mesoscale city-country breeze, which is also sometimes referred to as UHI flow. Under anticyclonic (e.g. warm, high pressure) conditions, this mesoscale system develops as thermals at the core of the city rise to the UBL generating advection flow at canopy-layer level from the cooler surroundings of the greenbelt (figure 6.13). As such, urban growth strategies that encroach in to peripheral green space can reduce this beneficial breeze by reducing the city-country temperature gradient, and hence the potential of the system, preventing the supply of relatively cooler air from the greenbelt vegetation. In contrast, compaction strategies and urban densification actually enhance this phenomenon by enhancing the city-country temperature gradient. The amount of cooling expected or lost due to loss of greenbelt depends upon the type and spread of the vegetation lost.

Until recently, reduced evapotranspiration in cities was considered the dominant factor for the formation of the daytime UHI. However, studies of US cities

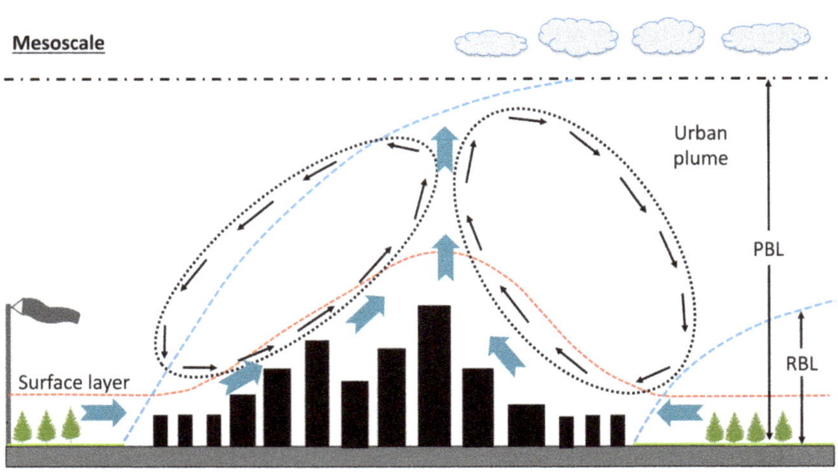

Figure 6.13. Schematic of the city-country breeze phenomenon.

demonstrated that the daytime UHI is primarily dependent on the relative effectiveness with which urban and rural areas convect heat to the atmosphere, rather than evaporative cooling (heat storage is the dominant factor for the night-time UHI). These studies found that the magnitude of the UHI was not directly correlated with precipitation or calculated potential evapotranspiration, but rather with convection efficiency. This study suggested that if urban areas are aerodynamically smoother than the surrounding rural areas (due to dense vegetation), then heat dissipation is less efficient resulting in warming, and if the opposite is true, then a cooling effect can be expected. The variation in convection efficiency between urban and rural locations is thought to be dependent on the background climate and its impact on vegetation cover in rural areas. In humid temperate climates in the US, convection was found to be less efficient at dissipating heat from urban areas than from rural areas because rural areas tended to be aerodynamically rougher than the urban areas due to the presence of denser and coarser vegetation canopies. This study found that US cities in such humid temperate climates showed a 58% reduction in convection efficiency compared to adjacent rural areas, resulting in a temperature increase of $\sim$3 °C during the daytime. The opposite was found, however, for US cities in drier climates because the urban area tended to be aerodynamically rougher than the surrounding rural landscape, where the drier conditions inhibit the growth of dense vegetation. In such US cities, a 1.5 °C decrease in daytime UHI intensity was observed. This study concluded, based upon climate model data and supporting field measurements, that the primary driver for the daytime UHI is the relative convection efficiency of the urban area compared to its surroundings rather than the reduction of evapotranspiration.

The formation of wind systems plays a significant role in the distribution of cooling from vegetated spaces. Macro-to-mesoscale prevailing wind flow and direction over the city determines the downwind spread, aided by a combination of simple advection along aligned street canyons and the turbulent mixing above the rooftops of cross canyons. The formation of microscale wind systems has also been identified to play a significant role in the horizontal distribution of cooling from urban green spaces in a similar way to that from peripheral greenbelt. Under anticyclonic conditions, thermals rising from the surrounding urban area generates low-level advection currents that draw air from the park as 'park-breezes'. This park-breeze effect generates a centripetal convection current, which completes its cycle with the subsidence of warmer urban air from above into the park (see figure 6.14).

The presence of a park-breeze system may well explain why cooling rates observed within urban parks are seldom comparable to those of rural areas and are more closely associated with that of the surrounding urban area. The park-breeze phenomenon can also explain why many parks do not appear on heat island intensity maps, such as figure 6.11. The presence of such a centripetal convection current hinders the vertical transport of cooling from the green space. The park-breeze system will only occur during periods of dynamic stability because higher wind velocities (>5 m s^{-1}) will create turbulent mixing and will disrupt the buoyancy effects. The low wind velocities typical of anticyclonic conditions, heatwaves and high UHI intensity favours the formation of such a buoyancy-driven centripetal

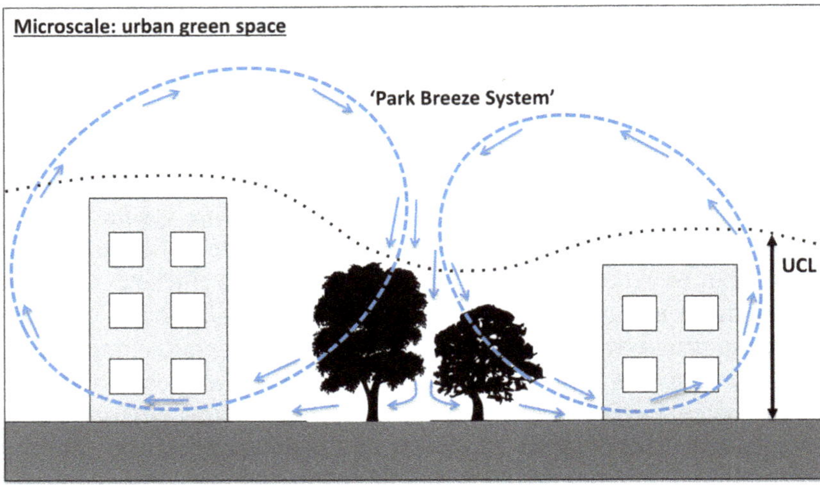

Figure 6.14. Schematic of the park-breeze phenomenon.

convection current. This implies that the UCL cooling influence of green space is enhanced by such microscale processes, offering greater cooling distribution when it is most needed to relieve heat stress.

Studies have shown that cooling from green space penetrates further into the surrounding urban context for larger parks. The significance of scale can be attributed to the increased potential of the park-breeze system, either due to an increased temperature gradient or else due to increased fetch (length of area over which a given fluid flow has contact). The geometry of the park is important here, with square or round-shapes providing greater cooling efficiency and distribution. This can be explained by a greater opportunity for increased temperature and humidity gradients and fetch between the park and the surrounding urban context. For a regular shape, the fetch will be similar irrespective of wind direction making the formation of a park-breeze more likely. The range of the cooling distribution observed is also dependent on the types and diversity of vegetation. A recent modelling study combined tree age and planting density as a composite leaf area index as means to calculate the optimum cooling effect relative to park size. The results of this modelling supported the findings of previous studies in that networks of smaller 0.2–0.3 km^2 green spaces can provide effective cooling distribution. An earlier study that considered the scale and interval between bodies suggested that such network or cluster arrangements should be spaced <300 m apart in order to provide their combined benefit. There is, however, a minimum effective size of bodies to consider, green space smaller than 0.05 km^2 have been shown to offer negligible cooling. This reinforces the hypothesis that a certain fetch is required to generate a park-breeze system and that larger parks are able to create larger park-breezes, allowing for greater cooling transport into the surrounding urban area, even for a constant temperature gradient. Currently, it is uncertain if the same effect can be achieved through networks of smaller green-spaces, and what the required size and interval is in relation to surrounding urban surface roughness.

The findings of the academic studies summarised above suggest that the addition of vegetation to an urban area with the principal aim of increasing evapotranspiration to mitigate the daytime UHI will have a minimal effect. At a mesoscale, the presence of vegetation seems to only aid the cooling of the city by enhancing its surface roughness. This provides another insight into the results presented in figures 6.10 and 6.11, where Richmond Park and Wimbledon Common presents a pronounced UHI reduction effect in contrast to Hyde Park and Kensington Gardens, not only because they are larger in area but also as they are significantly rougher aerodynamically. It can be hypothesised that if urban greening is to be utilised for this purpose, then tree planting with increased diversity of species will provide greater surface roughness than planar greening such as grassy parks. The type of urban greening to be used, therefore, requires consideration, not only transpiration potential but also the resultant aerodynamic roughness produced by varied arrangements. Certain countries have developed planning processes that account for the relative environmental capital of different green space land cover. For example, the green area ratio implemented in Berlin (Germany) and adapted in Malmo (Sweden) assign weighting factors based on their relative climate mitigation potential.

With regards to compact of dispersed development we have a choice, do we increase the density of our cities with taller buildings enhancing surface roughness and increasing the city-country temperature gradient and rely upon the city-country breeze to provide UHI cooling? Or, do we have more dispersed developments with networks of green infrastructure to provide park-breeze UHI cooling? The former is more energy and infrastructure efficient, but the latter provides greater environmental capital. The best solution is currently unclear and each urban area needs to be considered independently, weighing all the competing environmental and market forces.

6.7 Green building envelopes

One solution to allow urban densification and compact urban growth and increase green space is the use of green building envelopes, which can take the form of vegetated façades or roofs. There is an increasing number of advocates for vegetal architecture (incorporating vegetation into architecture) who wish for trees and shrubs to be incorporated onto buildings as either retrofits or replacements. Particular examples of such approaches include the Park Royal Hotel on Pickering Avenue, Singapore, The Kensington Roof Garden, London and Bosco Verticale, Milan (Italy).

However, despite this pressure, it is unlikely that such greening will be possible in already compacted urban centres. A study of Hong Kong acknowledged that with extreme urban centre densities and limitations on roof load bearing capacities, only planar greening solutions such as herbaceous sward green-roofs may be possible. This study, however, argued that green-roofs are less effective than street level vegetation for UCL cooling, particularly when typical building height exceeds 10 m. This is particularly true for Hong Kong where the average building height is 60 m, as such the UHI cooling influence of green-roofs was deemed negligible. While the cooling effects of vegetated roofs will not permeate to street level and directly reduce the UHI, there are the added benefits of reducing the albedo of the urban surface and the reduced heat penetration into the top floors of buildings, which will indirectly reduce the UHI. This has to be weighed against cheaper options, however, such as just painting the roof white (so-called cool-roofs). Some studies have demonstrated that in comparison to extensive green-roofs (primarily grasses or Sedum type plants), cool-roofs may result in lower annual (heating and cooling) energy costs. Although when considered against the intensive green-roofs that includes larger vegetative cover and greater heterogeneity, the opposite may be expected. The overall energy saving has been shown to generally increase with increased vegetation LAI and in the case of cooling-dominated buildings it is highlighted as the critical parameter. In terms of economic comparison, cool-roofing strategies tend to offer better value than green-roofing, which typically has higher construction and maintenance costs. The use of such greening strategies needs to be assessed beyond their simple economic practicality and consider other benefits, such as thermal insulation, reduced rainwater runoff, carbon uptake and the ecosystem benefits offered.

A recent review of studies of green roofs concluded that proximity is significant to the cooling experienced and suggested that there is limited vertical transport of air, likely due to limited aerodynamic roughness (particularly for extensive roofs). Since the cooling from green-roofs does not affect temperatures in street canyons and there is little vertical transport of air to affect the mesoscale UHI, we can conclude that green roofs are an ineffective UHI reduction measure. This is not to say that green roofs do not have a place in a modern dense city given that they have many other benefits beyond the evapotranspirative cooling of air.

An alternative to green-roofs are green-façades. At first glance, these may seem to be more appropriate to urban cooling as they are located within a street canyon inside the UCL. However, studies of such vertical greening solutions found that their cooling influence did not extend beyond their immediate foliage zone. As such the principle cooling attribute is to shade building façades and reduce surface temperatures. There are, however, added benefits in that such vegetation can help remove air pollution and provide some cooling naturally ventilated buildings. For more intensive living-wall type green façades, which include irrigation systems and can support larger plants, the range of cooling influence is more evident, although still restricted to <1 m. The technology behind living walls is constantly evolving and few in situ measurements have been published, so making generalisations about this type of greening strategy is risky at this time. For instance, the Rubens Hotel in London incorporates a 350 m^2 soil-based living-wall that also acts as a vertical sustainable drainage system. The evaporative cooling effect of this system is likely measurable and is no doubt beneficial in cooling the local air, but how far this cooling penetrates into the surrounding urban context is unknown. The impact of vertical greening on street canyon conditions, where surface temperatures and heat storage determine the nocturnal canopy-layer temperatures, has not yet been sufficiently studied. The cooling potential and delivery of other ecosystem services in dense urban settings can be considered as an emerging field of research and we cannot at this time quantify their benefit, if any. As such, incorporation of green façades and living walls should be done on the basis of their better-known ecosystem benefits.

6.8 Thermal properties of blue space

Urban blue space refers to the general category of features that incorporate the presence of substantial bodies of static or dynamic surface water. In many cities, substantial blue spaces naturally exist as part of their geography (e.g. rivers). In London, for example, the River Thames is a dominant feature, which together with other blue space represents ~2.5% of the city's area. Like green space, blue space can also be created or managed specifically to provide additional ecosystem services and local amenities. These can include reservoirs, canals, sustainable drainage systems and rainwater harvesting systems. City planners and architects have long considered water bodies as a vital strategy in minimising urban heat stress. Studies focusing on the cooling benefit of blue spaces are fewer compared to those for greenspace and typically focus on the daytime influence on urban temperatures.

A water body's (static) or watercourse's (dynamic) ability to modify surrounding air temperatures is dependent on its inherent properties (temperature, area, depth, waviness etc) and its interaction with the surrounding climate. Most studies focus on the evaporative cooling ability of blue space, transforming sensible heat into latent heat through the production of water vapour. The thermal properties of high specific heat capacity and enthalpy of vaporisation (or latent heat) give water a high thermal inertia, allowing water bodies to act as a thermal buffer playing an important role in moderating temperatures and temporal variations. For large water bodies such as oceans, over 90% of the solar radiation received over a year may be used to evaporate water, and for smaller water bodies this energy conversion is still likely to be >50%. Although on average this translates to a lower Bowen ratio for smaller, shallower water bodies, the evaporative flux is characterised by diurnal variations. For a large part of the morning, the absorbed solar radiation is mostly used to warm the water (increase heat storage). Towards the afternoon when the water surface temperature and the water-to-air vapour pressure deficit peak, a strong evaporative flux is generated. The energy stored within the water body is adequate to maintain this evaporative flux even throughout the night, albeit with diminishing intensity.

The surface reflectance (albedo) of a water surface determines its radiative exchanges. The albedo of blue space is low (~0.1) at low to medium angles of incident solar radiation typical of higher latitudes, and varies day-to-day with flow rate and waviness, the biochemical makeup and the presence of suspended particles (turbidity). This means that the majority of incident shortwave solar radiation which includes ultraviolet, visible and near-infrared light is absorbed by the water leading to warming of the water body. In most water bodies, solar radiation cannot penetrate any further than ~10 m. As we discussed in Chapter 1, water vapour is a greenhouse gas and is opaque to thermal IR radiation and liquid water is almost a perfect blackbody in the thermal IR part of the spectrum. Therefore, longwave (or thermal) infrared (wavelengths 4–100 μm) radiative flux from the atmosphere or surroundings is almost completely absorbed at the surface with minimal reflection, while the outgoing longwave flux remains relatively constant throughout the day (for larger water bodies), owing to the large thermal inertia and the limited diurnal variation in surface water temperatures. The fluidic properties of water allow absorbed radiative energy to be transferred within the water body by conduction, internal radiation and convection currents, all of which contribute to the efficient heat transport and mixing within the water body. These processes allow heat to be efficiently diffused throughout a large surface volume, maintaining surface water temperatures within a small diurnal range.

Higher surface water temperatures enhance the water-to-atmosphere temperature gradient. The temperature of the air above the water surface determines the sensible heat flux, with warmer air suppressing radiative heat loss and cooler air enhancing the water-to-atmosphere temperature gradient and hence increasing radiative flux. The water-to-atmosphere moisture gradient, also called the vapour pressure deficit, determines the potential for moisture to transfer to the atmosphere. Relatively drier air above the water body enhances evaporative flux as the vapour pressure deficit is increased, while more humid air above water surface suppresses evaporative flux. Increasing the wind velocity above a water body significantly alters both the sensible and evaporative heat flux by advecting away heat and moisture, enhancing both the temperature and moisture gradients. As with green spaces, high wind velocities can hinder cooling distribution by enhancing atmospheric mixing.

The thermal properties of dynamic (moving) watercourses such as rivers, streams and canals are influenced by fluid flow variables (e.g. volume flow rate) and local climate parameters. The flow of water allows watercourses to carry absorbed radiation downstream by advection and release this energy in a different location, potentially outside the urban area. Measurements of watercourses have shown a marked increase in water temperatures downstream of urban areas, with temperatures shown to increase with increasing impervious surface area and surface runoff in the urban area. A study of urban streams in Long Island (USA) found water temperatures to be 5 °C–8 °C warmer in summer and 1.5 °C–3 °C cooler in winter than rural streams. This study also noted that the diurnal temperature variation was greater in urban streams. During the summer, stormwater runoff from heated impervious surfaces lead to increases in the temperature of the urban streams of 10 °C–15 °C compared to rural streams. Although such stormwater runoff has a beneficial cooling influence on the upstream urban surface, it can lead to thermal pollution and resulting biochemical issues, such as algae blooms further downstream.

The energy balance of a river is dominated by the net shortwave radiation balance, followed by contributions from the net longwave radiation balance and evaporative flux. For example, a study of River Exe in Devon (UK) showed the net radiation balance to account for 56% of the heat gain and 49% of heat loss. Since most watercourses are of limited depth for most of their course, thermal exchanges at the riverbed-water interface also require attention. Another study, this time of the River Blithe in Staffordshire (UK), found that 82% of energy exchange occurred at the air–water interface, while only ~15% occurred at the riverbed-water interface. For smaller streams, factors such as groundwater exchange and discharge from external sources have a greater influence on the total energy exchange, while for larger dynamic bodies, exposure to solar radiation and wind flow lead to the dominant heat exchange occurring at the air–water interface. This means that their water temperatures typically depend on the climatic conditions above and their diurnal and seasonal cycles.

The limited flow of static water bodies restricts dynamic instability, as such they demonstrate higher sensitivity to energy exchange modifications at the air–water interface, meaning that their thermal properties are highly dependent on the climatic conditions above. For deep static bodies such as large lakes and reservoirs, thermal inputs can lead to temperature and density changes that produce strong thermal stratification of the water column. In such stratified water bodies, the *epilimnion* describes the warmer upper layer, followed by the *metalimnion* or *thermocline* where the temperature rapidly decreases with depth and the cooler, denser and more stable *hypolimnion* at the bottom of the water body. This stratification of deeper water bodies means that the thermally active zone for energy exchange with the atmosphere above is restricted to the epilimnion and the higher regions of the thermocline. Several mixing mechanisms can affect the thermal moderation of this active zone. At the surface of a water body, evaporative mixing can be observed, evaporation is an endothermic process, hence the latent flux generates instability that brings warmer water to the surface to maintain a relatively constant surface temperature. Of far greater significance is the mechanical energy transferred by wind shear stresses at the water surface, producing waves and turbulence. The strength of this mixing is dependent on the characteristics of the wind flow at the water body's surface, as well as the available fetch and the presence of any littoral (riverbank/shoreline) obstructions. Strong wind-driven turbulence in the water body may transfer turbulent kinetic energy to the lower layers leading to a breakdown of the stratification. In addition to these forms of surface mixing, a diurnal mixing current can be generated at the littoral zone (near the banks) of a water body. Since the shallower water on the littoral slope heats faster than the open water during the day, a horizontal current is generated from the littoral zone towards open water, while cooler water from the open water depth is drawn up the slope. As the littoral zone cools faster than open waters at night, this current is reversed. This diurnal littoral zone current, however, is generally not strong enough to disrupt the stratification of the entire water body.

Whole water body mixing is evident in holomictic water bodies common to temperate climates, which have a uniform temperature and density from top to bottom during certain seasons. The seasonal changes in temperate climates produce the conditions necessary for buoyancy-driven overturning of the stratification structure. This overturning occurs at a threshold temperature of $\sim$4 °C, when water reaches its maximum density. During the spring, when surface water that is cooler than this threshold temperature starts to warm, its density increases and falls through the water body creating a convective instability. This continues until the epilimnion reaches $\sim$4 °C. After this point, further warming increases stability and restricts vertical mixing leading to stratification. The reverse is true in autumn, as the surface water cools, its density increases, again creating convective instability and the mixing of the whole water body. Water bodies that demonstrate this type of biannual mixing are referred to as dimictic bodies, and water bodies in cold winter climates have been observed to demonstrate rapid overturning. Apart from this seasonal overturning, all other natural mixing processes act to maintain the stratification structure for most of the year. During these periods, only the epilimnion and the upper thermocline act in energy exchange with the climate above because the stratified state of the water body prevents the use of its entire volume and thermal capacity.

For shallower static bodies, their limited volume places a limit on the available thermal capacity and inertia. As such, the surface temperature and resulting latent flux peaks earlier (daily and seasonally) than for deeper body. The reduced volume of water not only increases the significance of processes discussed above, such as littoral zone and wind-driven mixing, but the conduction of heat across the water-bank boundary into the surrounding ground and the absorption of heat by aquatic flora and fauna are all likely to be more pronounced. The increased effect of these energetic processes means that for shallower water bodies the net radiation balance converted into evaporative flux is lower than for larger ones, but still a substantial value. For instance, a study of a shallow lake in the Hudson Bay lowlands (Canada) showed that on average 55% of the daily net radiation balance was still utilised for evaporative flux. The reduced depth of such water bodies allows shortwave radiation to penetration the full depth, allowing solar radiation to be absorbed (or reflected back) and heat to be conducted into the subsurface. This can lead to water-to-bed interface energy exchanges and possible warming of the water from below. Shallow water bodies are generally considered not to exhibit stratification of the water column due to the wind-driven and evaporative mixing, instead shallow water bodies are often considered to be a well-mixed epilimnion, able to use their entire volume and thermal capacity for energy exchange. However, there is evidence that even shallow water bodies can demonstrate stratification during prolonged periods of warm calm weather typical of heatwaves and high UHI intensity, which reduces their effective thermal capacity and limits air–water energy exchange when it would be most beneficial.

Smaller and shallower ponds have become more commonplace as mechanisms for sustainable (urban) drainage system (SuDS), which is now common practice within urban development plans. The urban setting has a profound effect on the thermal properties of such ponds. The inflow of warm summertime surface runoff as well wind sheltering from surrounding buildings inhibiting wind-driven surface mixing means that urban ponds behave different thermally from their rural counterparts. These urban influences may result in the development of thermal stratification in such ponds. For example, a study of 10 shallow urban ponds in Ontario (Canada) showed that the density variation produced by daytime heating was not always fully dissipated at night depending on the surrounding environmental and subsurface conditions. The resultant stability and stratification extended through the mid-summer months, with the ponds demonstrating vertical temperature differences of ~4 °C between the top and bottom layers. This strong and persistent stratification in a shallow pond was attributed to the turbidity of the water, which contained high levels of suspended sediment and organic particles, and the reduced wind stress observed at the water surface.

In terms of geometric parameters, the ratio between the water's surface area and perimeter (shape factor) and the maximum depth are significant. Examination of the influence of shape factor reveals that shallow ponds with relatively large area but with simple a geometry (round or square) experience pronounced stratification as opposed to the condition in larger lakes where larger surface areas with greater fetch are typically associated with greater turbulent mixing. The maximum depth unsurprisingly demonstrates the greatest correlation with stratification, with only the shallowest water bodies <1 m deep identified as being relatively isothermal (a well-mixed epilimnion). This means that only very shallow water bodies are able to utilise their entire volumes thermal capacity for air–water energy exchange.

6.9 Thermal effects of blue space

As with green spaces, the cooling effectiveness of blue space in terms of magnitude and distribution is dependent amongst other things by the size and spread of bodies and the distance from them. A recent mesoscale modelling study of a hypothetical water bodies situated within an idealised city showed that relatively large water bodies demonstrate greater cooling, particularly close to their boundaries and in downwind areas. The length of the downwind distribution was dependent on wind velocity, with relatively cooler air from a large urban water body bale to create plumes several km long. This modelling study also confirmed previous findings that several smaller regularly shaped water bodies distributed evenly throughout the urban area have a smaller cooling effect (particularly during the day), albeit across a greater area.

A study of Beijing (China) observed that water body geometry is significant for cooling distribution, with square or round geometries providing greater cooling efficiency. As discussed previously in relation to green space, this can be attributed to the increased temperature and moisture gradients that are likely to result between such wider-shaped water bodies and their surrounding landscape. Additionally, regular geometries present a more consistent fetch for wind irrespective of wind direction, presenting greater opportunities for atmospheric advection. Similarly, the Beijing study highlighted the importance of a watercourses width, with urban river width being a principal factor affecting the temperature and humidity of the riparian (area adjacent to rivers) zone. This study found that when a river is >40 m wide, there is a significant and stable effect of decreasing temperatures and increasing humidity, indicating the transport of air away from the water's surface. This may explain why the River Lea in the Lee Valley Regional Park, which is <30 m wide through much of the park, does not appear to have a cooling effect on the atmospheric UHI for London shown earlier in figure 6.11.

An urban setting modifies the climatic variables that affect a water body; for instance, higher water surface temperatures are observed for water bodies surrounded by buildings which store and radiate heat. This creates a greater temperature gradient between the centre of the water body and its surrounding context, enhancing cooling distribution. The surrounding urban morphology further influences this cooling distribution by shading the water body and obstructing wind flow. Shading alters the net radiation balance reducing the temperature gradient and evaporative flux, while wind sheltering reduces the opportunity for atmospheric advection and surface layer mixing from waves. Littoral sheltering, whether from buildings or vegetation, reduces the wind stress induced turbulent mixing (as does aquatic vegetation), resulting in a reduced surface layer depth (and reduced volume or energy exchange) and potential strengthening of temperature stratification. The degree of mixing generated depends on both the amount (cover) of sheltering present and the available fetch. The longer the fetch, the greater the opportunity for turbulent mixing. For larger lakes with increased fetch, littoral sheltering damps surface waves, preventing turbulent kinetic energy from penetrating into the water column. The effectiveness of this damping is dependent on obstruction separation, with higher obstruction density achieving greater damping. Meanwhile, as obstruction separation is increased, turbulence also increases until it is similar to open water.

Farms in Australia often have such obstructions upwind of small reservoirs to reduce the evaporative water loss. A study of the effect of these littoral windbreaks (both artificial and natural) on the evaporation from these small water bodies showed that reduced wind speeds and increased air turbulence immediately downwind of a shelter lead to the accumulation of moisture in the air immediately above the water surface, reducing the humidity gradient and evaporative flux. The area immediately downwind of the littoral obstruction showed reduced waviness of the water surface, and hence the water's surface area in contact with the atmosphere is also reduced, this is termed the *quiet zone*. Wind speeds downwind of the obstruction decrease to a minimum at ~3× the height of the obstruction (see figure 6.15). While the location minimum wind speed does not mark the edge of the quiet zone, the two are related, with:

$$x\_\text{quiet} = 1.5 + 0.92x\_\text{min}$$

Wind speeds (u) further downwind of the obstruction remain reduced, although turbulent intensity increases as wind flow recovers to that upwind of the obstruction. This turbulent *wake zone* exhibits increased moisture and heat transfer. This study found that greater wind protection reduces the wind speed experienced by the water body, and therefore reduces evaporative cooling; however, a corresponding increase in surface water temperature was not observed. It can be hypothesised that this is due to increased radiative output, or increased conductive heat loss between the water surface and the moisture saturated air that forms above the water surface. Finally, this study found that evaporation from the water's surface varied throughout the day with a distinct diurnal cycle. At night when the Richardson number (the ratio between buoyancy and flow shear or equivalently the ratio between natural and forced convection) was suggestive of atmospheric instability (see figure 6.6), the sheltering effect of a windbreak was reduced.

The sheltering effect of natural (e.g. trees) and artificial (e.g. buildings) littoral windbreaks is another possible reason why Lee Valley Park which contains several lakes and reservoirs does not appear to affect the UHI shown in figure 6.11. Due to the linear geometry of the park and surrounding trees, the water surface may well be completely sheltered (i.e. entirely within the quiet zone), reducing the cooling influence of the blue spaces. Even within the wake zone, where evaporation is increased due to air turbulence, the same turbulence will inhibit the vertical

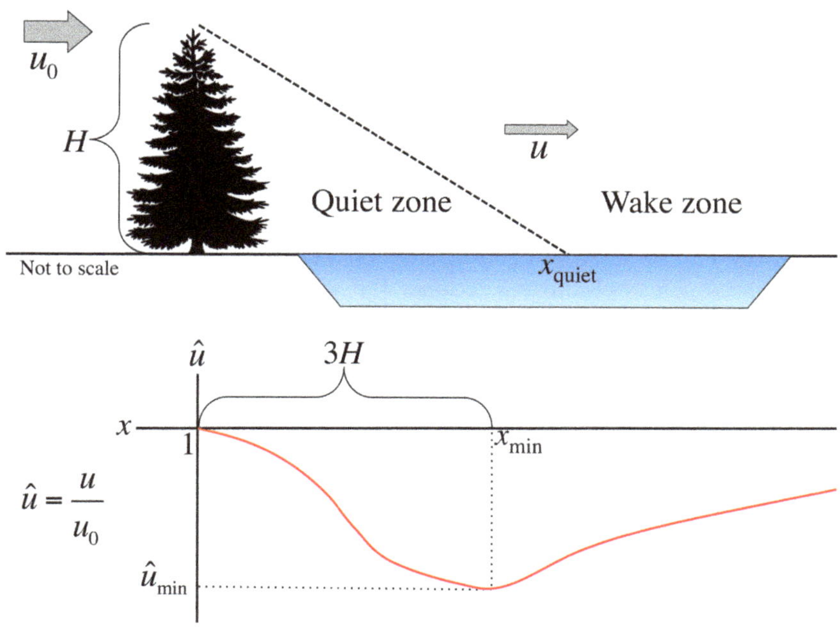

Figure 6.15. Sheltering effects of littoral obstructions on a water body and resultant wind velocities.

transport of cooling. This implies that thought has to be given to the placement of blue space in relation to the prevailing wind direction and any surrounding littoral windbreaks to avoid creating localised stagnant areas that exhibit higher night-time temperatures (due to the large thermal inertia of water) and higher humidity (from moisture saturated air), which can be damaging to human health.

Blue space cooling distribution demonstrates distinct diurnal and seasonal variations. For example, a study of an urban river in Sheffield (UK) observed that the cooling effect tends to be greatest in the morning, with warm days in spring demonstrating ~2 °C lower temperatures over the river and ~1.5 °C in the riparian zone. No night-time cooling was observed. Towards late-June (mid-summer), daytime cooling had notably diminished with similar air temperatures observed over the water and elsewhere. In agreement with such observations, the mesoscale model of a hypothetical city mentioned above showed blue space cooling as principally a daytime phenomenon, while at night and particularly towards the mid/end of the summer, a distinct nocturnal warming effect can be expected.

These diurnal and seasonal variations can be explained by the variation in the evaporative flux. The instability of the atmosphere in the morning will increase the evaporative flux into the immediate atmosphere above the water body, increasing its water vapour content. In the early afternoon, atmospheric convective instability peaks along with the water surface temperature and evaporative flux, although the vapour content of the air immediate above the water body reduces as the surface warms, increasing the buoyancy of the water vapour transporting it to higher altitudes where concentrations are diluted. The convective instability of the atmosphere reduces during the day and in the evening and at night as the surface atmosphere cools gains stability and resistance to vertical transport (see figure 5.6). The stabilising of the atmosphere reduces the ability to transport water vapour to higher altitudes, as such evaporation from the warm waters surface during the evening leads to increasing moisture content near the surface. Hence, the stable layers immediately above the water surface become saturated with evaporated water vapour, reducing the moisture gradient and preventing any further evaporative flux.

The reduction in evaporative flux during the night results in a warming of the water's surface. The differential cooling rates of the water body and its surrounding urban context (which cools faster) reduces the water body-to-context temperature gradient. This reduces the likelihood of nocturnal horizontal advection currents, which could vacate the saturated stable air mass above the water body. This warming effect is particularly pronounced when water bodies reach higher temperatures towards the end of the summer from accumulated thermal energy. Simulations of the diurnal cycle of a water body indicate a cycle of warming and cooling, as follows:

- Morning (~10 am): cooling at low altitudes (<500 m), warming at higher altitudes.
- Afternoon (~3 pm): modest cooling at all altitudes, greatest vertical transport.
- Night-time (~4 am): warming but at low altitudes only, minimal vertical transport.

It is possible to hypothesise that differing thermal properties of water bodies and the surrounding urban context generate distinct microscale breeze systems, similar to those observed above green space. Such a water body-breeze system would differ from a park-breeze system due to the high thermal inertia of water and its diurnal cycle. The water body-breeze system could thus be expected to reverse during the night as warm saturated air rises from the warmer water body causing cooler air from the urban surroundings to advect towards the water body (figure 6.16). The completion of this

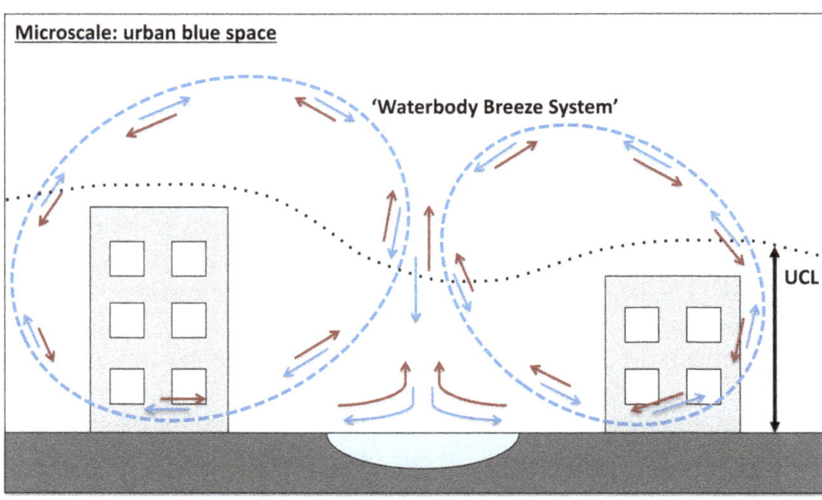

Figure 6.16. Schematic of the water body-breeze phenomenon.

water body-breeze centripetal cycle would be the night-time subsidence of warmer and humid air back to the surrounding urban area. This presents the possibility for the horizontal transport of an undesirable warming effect into the surrounding areas. The canopy-layer trapping of heat and moisture presents a significant threat to not only thermal comfort but also to human health and wellbeing.

The suggested cooling benefit of evaporation from water bodies is somewhat misleading, even during the day. A significant drawback of evaporative cooling is that it increases the water vapour content of the air (humidity). Increased humidity reduces the effectiveness of human thermoregulation by reducing the readiness with which sweat can evaporate. In addition, since water vapour is a greenhouse gas by altering the emissivity of the surrounding air leading to greater absorption and reradiation, it traps heat at street level. Studies have observed that up ~60% of the cooling benefit from reduced air temperatures (sensible cooling) can be negated by the increase in humidity. Consideration of the diurnal thermal cycle of blue space suggests that it warms the urban environment when it is least desirable: at night and particularly under anticyclonic conditions that are typical of high UHI intensity and heatwaves. Hence, we can conclude that blue space offers limited potential for UHI mitigation if considered in isolation.

With so much of the thermal effects of water bodies in urban areas still unknown, this is an active area of research. The evaporation of water is an endothermic process which should produce cooler water vapour, and water vapour being less dense than air should provide vertical transport of heat, air and pollutants away from the urban surface. However, there are competing factors, the boundary layer produced by the cityscape will act to limit vertical transport. In addition, the thermal mass of the water will alter the radiative exchange within the street canyons during the day and act as a source of heating during the evening, limiting the rate at which street canyons can cool. Add to this that evaporation may well increase street level humidity levels and uncertainty about if and when urban water bodies may stratify, limiting the volume of thermally active water, it is hardly surprising there is little agreement on how and whether blue spaces relieve heat stress in hot conditions. What can be certain is that the environmental conditions and level of atmospheric stability above the water body, both of which can change over the course of a day, will have an effect upon its thermal interactions, some indicative scenarios are shown in figure 6.17.

In an attempt to better understand how a water body behaves over the course of a day, Ampazidis *et al* [5, 6] used a computational fluid dynamics (CFD) software package to study the evaporation effects of water bodies under different conditions within an idealised urban area, consisting of a 7 × 3 array of buildings with the central building replaced by a water body. Conditions were varied to produce different convection regimes, namely forced convection (convection driven by fluid movement i.e. wind) and mixed convection (both natural, temperature/buoyancy driven and forced convection are present) conditions, which are both common within an urban area. These simulation conditions can be related to real world conditions through the use of the dimensionless Richardson number (Ri), where forced convection $Ri = \pm 0.04$ and mixed convection $Ri = \pm 1.1$ depending upon the direction of the convective motion defined by the temperature difference. The temperature

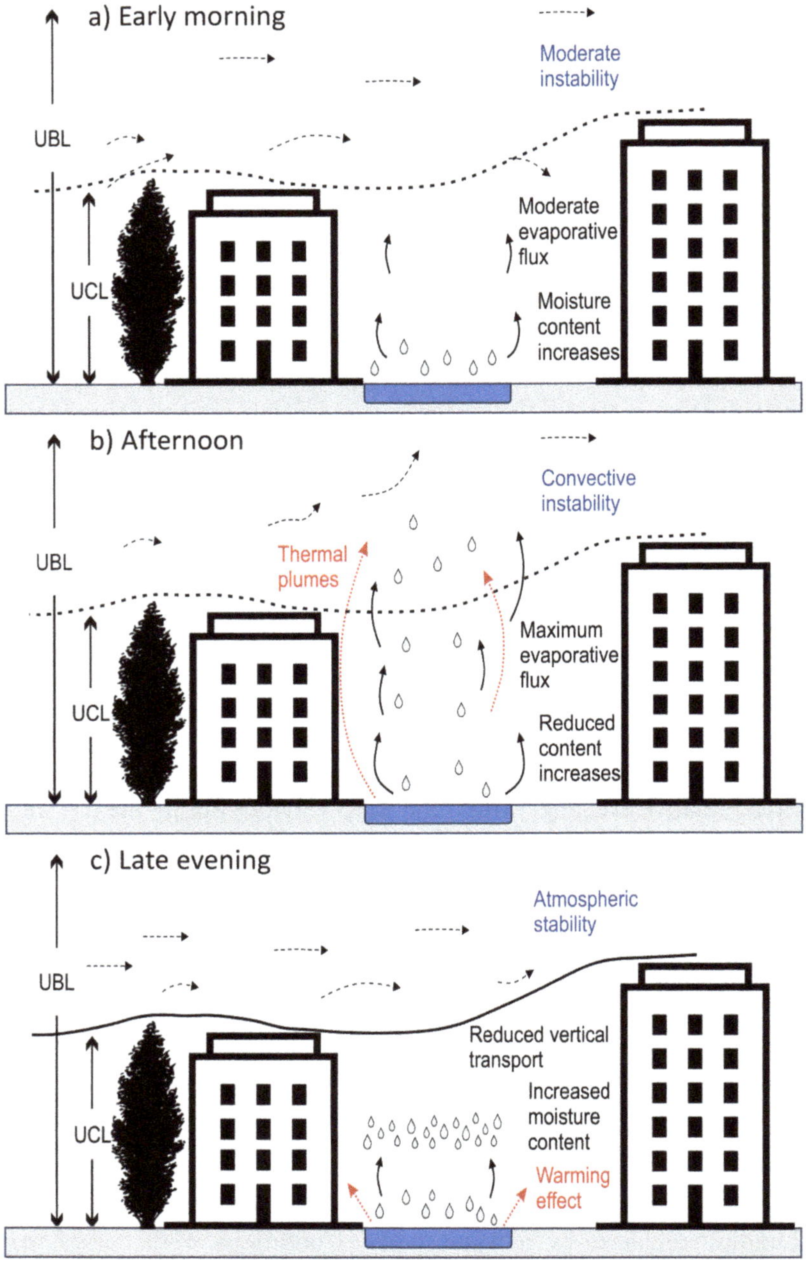

Figure 6.17. Diurnal evaporative cycle of an urban blue space. Note the different levels of atmospheric stability over the day. Adapted from [4], Copyright (2020), with permission from Elsevier.

difference between the water body and the background air was varied to approximate different times of day, such as a cooler water body for morning/daytime conditions, and a warmer water body to represent evening/overnight conditions.

Figure 6.18 shows the normalised air velocity streamlines (actual velocity divided by the initial velocity) for the different cases investigated by Ampatzidis *et al* [5]. What the simulations showed was that in the base case with no water body present in the central open square, flow is dominated by a large principal vortex induced by atmospheric wind flow and the building induced surface roughness. This form of circulatory flow within the open square acts to isolate the air at street level from the air above the tops of the buildings. We can see that for the forced convection cases ($Ri = \pm 0.4$), the dominance of the inertial forces from the increased wind speed over the natural buoyancy effects means that the water body has a minimal effect on the air flow within the open square, regardless of whether it is warmer or cooler than the surrounding air. Conversely, for the mixed convection cases ($Ri = \pm 1.1$), we see a distinct difference between the cooler and warmer water bodies. The baseline case under mixed convection (not shown) displays results very similar to the baseline case under forced convection (panel (a)); therefore, we can attribute the changes in flow purely to the presence of the water body and its production of water vapour. Panel (d) shows that the downward convection caused by the cooler water body ($Ri = -1.1$), creates an overturning of air within the open square, mainly in the upwind area on the right-hand side. This flow of cool air and water vapour disrupts the principal vortex that forms there under the baseline and forced convection cases. In the case of the warmer water body (panel (e)), the additional buoyancy created by warm water vapour induces a strong upwards motion of air and water vapour away from the

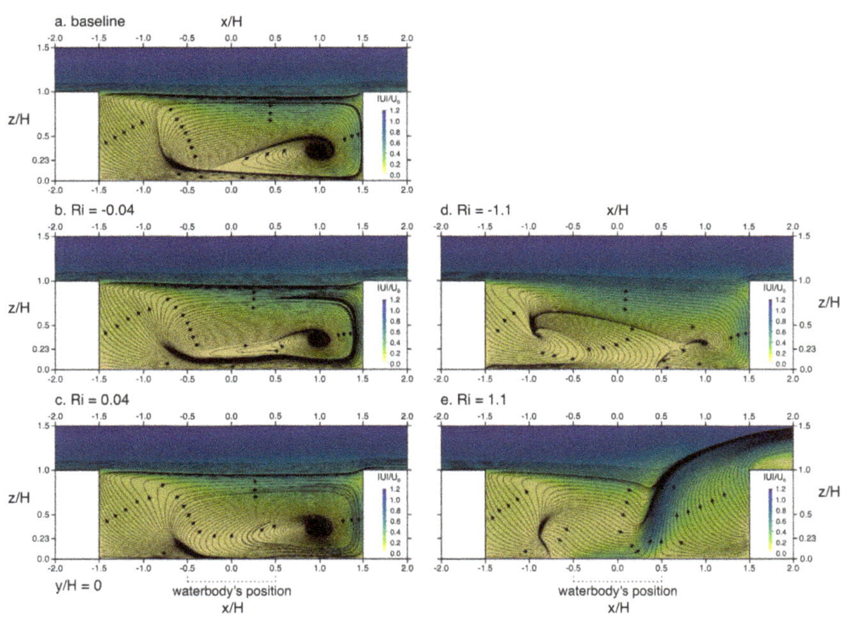

Figure 6.18. Normalised velocity streamlines on a vertical plane for a water body under different conditions. Note the position of the water body. High level wind flow is from left to right. Panel (a) is the base case with no water body. Panels (b) and (c) are a cooler and warmer water body, respectively, under forced convection conditions. Panels (d) and (e) are for a cooler and warmer water body respectively under mixed convection conditions. Reprinted from [5], Copyright (2022), with permission from Elsevier.

surface. Critically, here the upwards plume extends beyond the roof level meaning that air is leaving the open square and replacement air will be drawn in at street level from adjacent streets to replace it.

The results presented by Ampatzidis *et al* [5] are in general agreement with the findings of wind tunnel measurements of flows over scaled models, where the heated surfaces of street canyons promote similar vertical movement and the principal vortex breaks up with increasing Richardson numbers (e.g. more buoyancy from higher temperatures or slower airflow). In these cases, the principal vortex is destroyed and the vertical motion of the air extends downwind beyond the street canyon.

Figure 6.19 shows the effect of the water body on flow over the downwind street canyons. In the baseline case with no water body present in the open square, the downwind street canyons exhibit a single large vortex and are isolated from the airflow above the roof line. This situation persists for the other scenarios where water body does not create sufficient vertical motion to disrupt the skimming flow above. However, in the case of the warmer water body under the mixed convection regime (Ri=1.1), the plume created in the central open square sufficiently disrupts the skimming flow above the roof line and air from the downwind street canyons is drawn upwards. This is of great importance because it demonstrates that a single water body is able to remove heat and pollutants from the surrounding streets as well as just the local vicinity. This, however, only occurs in the Ri=1.1 scenario of the warmer water body and mixed convection regime, which can be considered typical of late afternoon/evening conditions (or equivalently late summer/autumn conditions). In the case of a cooler water body under a mixed convection regime

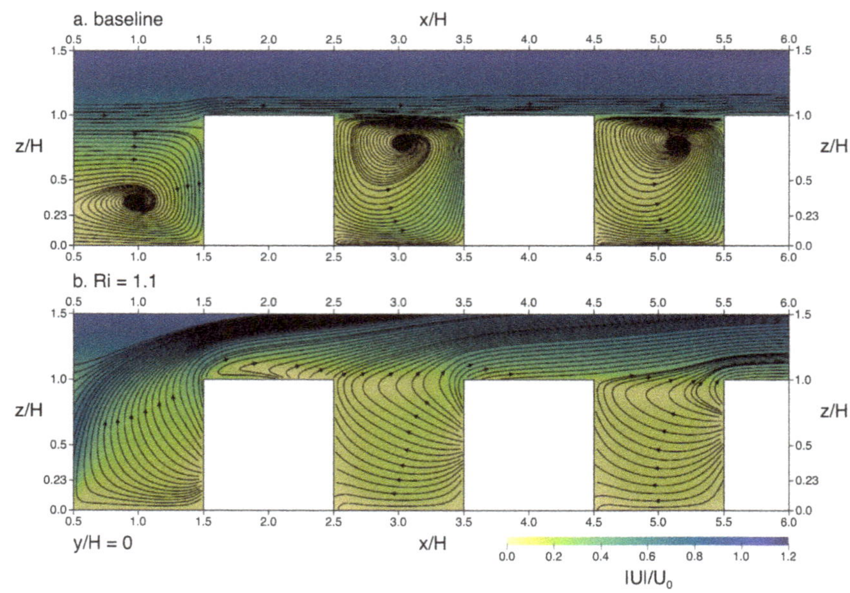

Figure 6.19. Normalised velocity streamlines on a vertical plane for the downwind street canyons under the mixed convection conditions. Panel (a) is the baseline scenario with no water body, while panel (b) is for a warmer water body with Ri=1.1. Reprinted from [5], Copyright (2022), with permission from Elsevier.

(Ri=−1.1), which could be considered typical of morning and early afternoon conditions (or equivalently springtime/early summer conditions), the influence of the water body is unable to break through the skimming flow above the roof line. In this case, the water vapour created by the evaporation from the water body is trapped at street level. This is also an important consideration because it means that during the hottest part of the day, a water body, which due to its thermal inertia lags behind air temperatures (i.e. is cooler), is not creating vertical movement of air to transport heat away from the surface. Furthermore, it is a source of humidity at street level, which will exacerbate heat stress and thermal discomfort. Figure 6.20 shows the

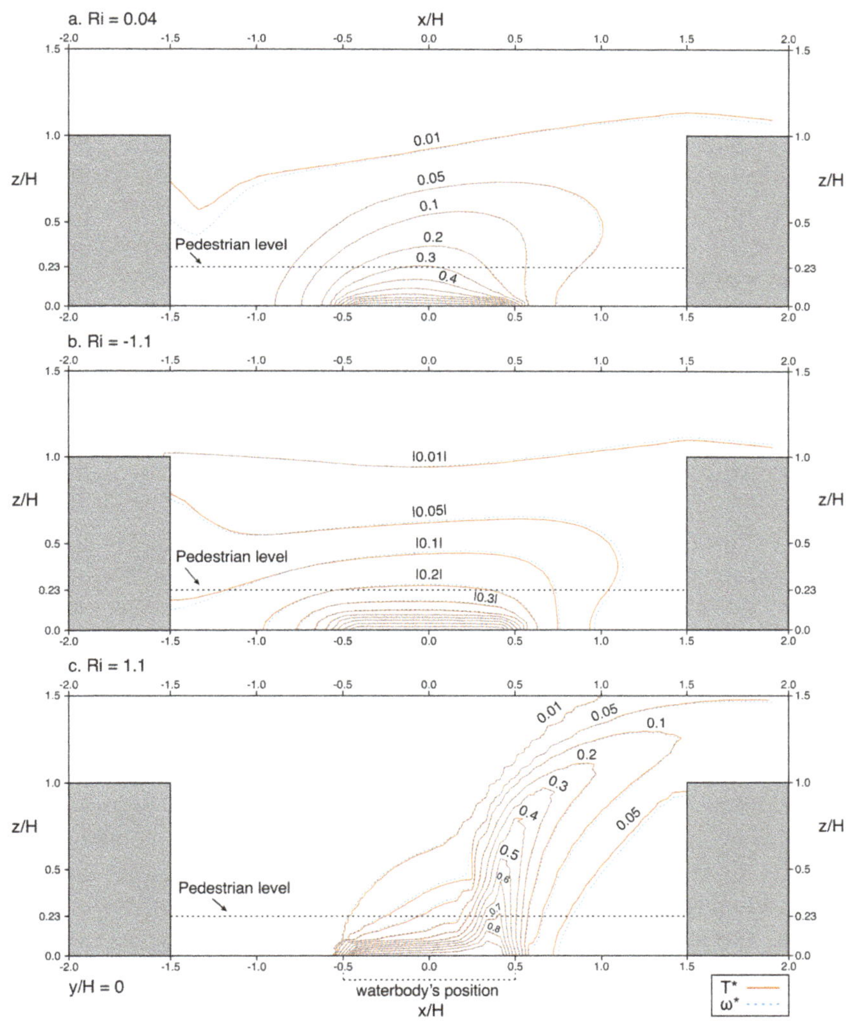

Figure 6.20. Contours of normalised temperature and water vapour concentration in the open square above the water body. Panel (a) shows the forced convection regime for a warmer water body (cooler not shown), while panels (b) and (c) show the mixed convection cases for a cooler and warmer water body, respectively. Reprinted from [5], Copyright (2022), with permission from Elsevier.

normalised temperature and water vapour concentration contours for the forced convections cases (warmer and cooler water body produce nearly identical results here) and also the mixed convection cases with a cooler water body in panel (b) and the warmer water body in panel (c). We can see that the warmer water body promotes an upward flow of heat and water vapour. However, the cooler water body sees the accumulation of cooler but more humid air throughout the open square surrounding the water body. The increased humidity will offset the decrease in temperature when it comes to human thermal comfort and may actually make conditions worse because the increased humidity will reduce the ability of the human body to cool itself via sweating. The forced convection regime shown in panel (a) does not suffer from the same issues because the vortices within the open square contain the water vapour largely to a region directly above the water surface away from pedestrians.

The work of Ampatzidis *et al* [5] has illustrated how the behaviour of an urban water body may change over the course of a day. This study demonstrated the effects of an area of blue space upon airflow air temperature and water vapour concentration distributions in a simplified urban area. While the different temperature and convection regimes allow an insight into the diurnal performance of an urban water body, all the simulations were performed under neutral atmospheric conditions. As such, we can assume that as the atmosphere stabilises over the course of the night-time (see figure 6.17), the strength of the boundary layer above roof level will increase and the increased buoyancy of the warm water vapour may become insufficient to break through and promote the vertical transport of heat, air and pollutants away from the surface. Under these conditions, we can hypothesise that the water body, which will likely still be warmer than the surrounding air, will continue to evaporate but that the warm water vapour will be contained at street level. This will dramatically worsen thermal comfort conditions and have a considerable impact upon dwellings trying to remove accumulated heat via natural ventilation (opening windows). A stable atmosphere is not only observed overnight but also during hot still weather, such as during heatwaves. It is clear, therefore, that significantly more work needs to be done to further explore the performance of urban water bodies under different conditions, so that blue infrastructure can be safely deployed into an urban area without risk of worsening environmental conditions and endangering human life during heatwaves.

Further work by Ampatzidis *et al* [6] examined the impact of the size and shape of a water body on its cooling potential. Using the same numerical solver, simulations were performed for both warmer and cooler water bodies of varying size from 1:1 (water body same size as the buildings in a regular array as used previously) down to 1:16 of the original size, as well as for progressively elongated water bodies of equivalent area. Again, a 7×3 array of buildings was used with the central building removed and a water body inserted in its place. Here, we will focus upon the effects of changing size for the warmer water body only. Figure 6.21 shows the results obtained for a 2 °C warmer water body under a mixed convection regime (Ri=1.9) in the central square, while figure 6.22 shows the results for the downwind street canyons. These are directly comparable with the data shown in figures 6.18 and 6.19.

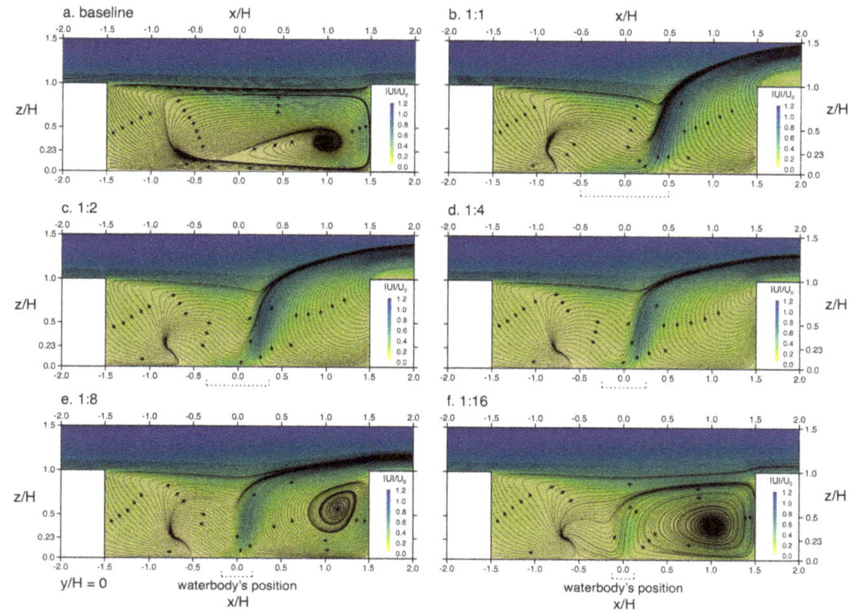

Figure 6.21. Normalised velocity streamlines on a vertical plane for different size water bodies. Note position of the water body. High level wind flow is from left to right. Panel (a) is the base case with no water body. Panels (b)–(f) depict progressively smaller water bodies. Reproduced from [6]. CC-BY-4.0.

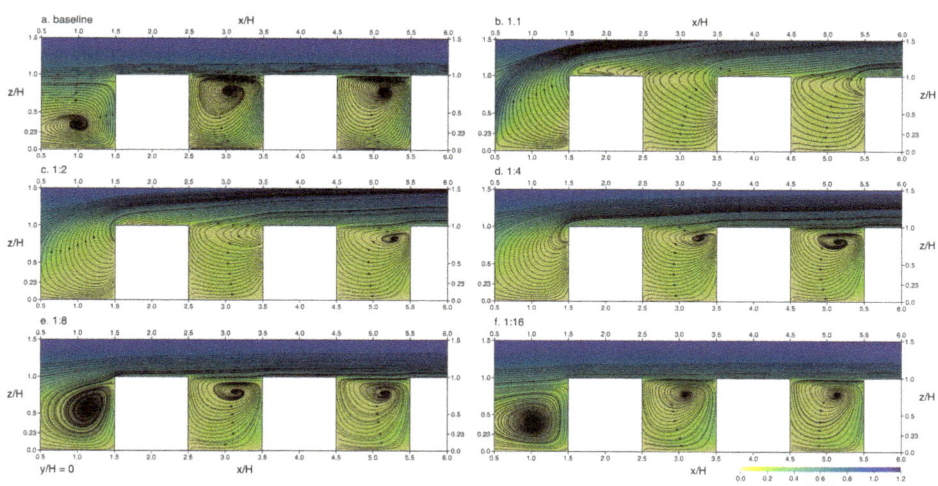

Figure 6.22. Normalised velocity streamlines on a vertical plane for the downwind street canyons for different size water bodies. Reproduced from [6]. CC-BY-4.0.

The results indicate that as a water body is made smaller, the amount of buoyancy is decreased and that inadequately sized water bodies are unable to promote sufficient vertical transport to break through the skimming layer above the roof line. The 1:16 case is completely isolated from the atmosphere above in the same way as the baseline where there is no water body present. This reducing buoyancy leads to overturning and increased temperature and humidity levels at street level, which increases the risk of heat stress and thermal discomfort. We can also see that as the size of the water body is reduced, the influence on the downwind street canyons is also diminished, with vortices starting to form indicating increasing isolation from the 1:2 case. In the 1:8 and 1:16 cases, the downwind street canyons are completely isolated from the atmosphere above with the water body having no influence on the airflow in these streets anymore.

We can conclude, therefore, that larger water bodies are better suited to the evening/night-time transport of heat and pollutants away from the urban surface. These results indicate that during late summer nights when the water is warmer than the surrounding air, an inadequately sized water body will promote overturning of the air within the local vicinity increasing temperatures and humidities at street level. The vertical transport will also be diminished during heatwave events with increased atmospheric stability. As such, if a water body is not adequately sized, then it can pose a real risk to human health during adverse heat conditions, which are predicted to become more common as a result of climate change. However, we should note that the potential of blue infrastructure solutions to mitigate the UHI and cool the urban environment and transport pollutants away from street level is considerable, especially when considering that blue infrastructure solutions can also be integrated into SuDS and can provide valuable amenities to the local populous.

6.10 Urban planning for the UHI

The design of urban areas, particularly those in warmer locations, has evolved over time to reduce the UHI, citywide temperatures and heat stress. This largely takes the form of buildings providing mutual shading or the alignment of street canyons to allow cooling breezes to penetrate deep into the city. With freely available satellite images, it is now relatively easy to compare urban layouts for different cities around the world. Figure 6.23 shows real street (red lines) and building (solid black lines) layout for sections of four cities at the same scale using data obtained from ArcGIS24. The buildings shown are primarily domestic to allow a fair comparison.

We can see from figure 6.23 that for cool maritime climates such as London and Bath, buildings have a compact form, typically with small openings to minimise heat loss. For warm dry climates such as that of Valencia, we can see from figure 6.23 that the urban form is more compact, buildings are closer together to provide mutual shading at the lower latitude, with central courtyards used to create a local microclimate, a well of cool air, to cool the inside of the building. As such, the urban street canyons have a greater height to width ratio than in cooler climates, reducing the UHI by limiting solar gain during the day. Buildings are typically light coloured to reflect more shortwave solar radiation. We can see from figure 6.23 that the urban form of Singapore is sparse with free spaces between buildings to promote airflow through the urban area. Another distinguishing feature between the urban form of the four cities shown in figure 6.23 is the amount and placement of vegetation. For the areas of London and Bath shown, there is some grass and a few trees located

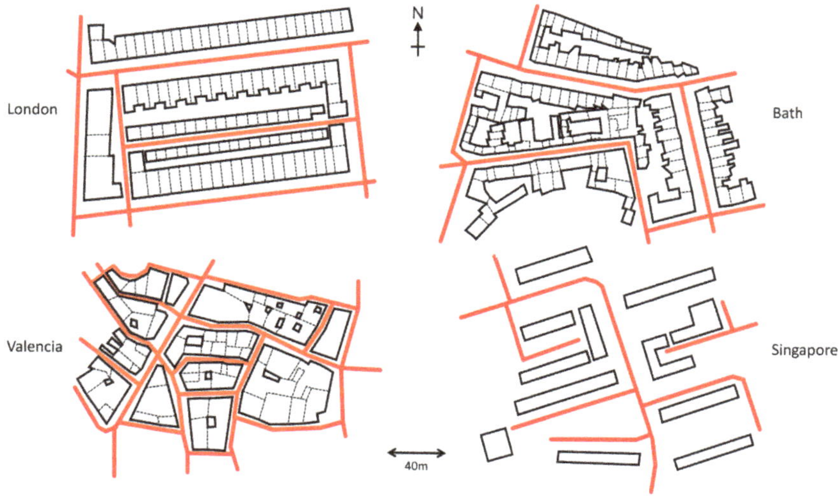

Figure 6.23. Example urban layouts for different cities, showing buildings (black) and roads (red) at the same scale.

in the back gardens of the homes. For Valencia, there is no grass visible in the satellite image but there is a greater number of trees, which line the roads and shade the façades in the street canyons. By comparison, satellite imagery of the area of Singapore shown contains more green space than man-made surfaces. There are a large number of trees with grass beneath, between the buildings and shading the façades. The large amount of vegetation in this urban area and the sparse urban form effectively minimises the UHI. These are just some of the architectural traits that have evolved over time in different locations to control temperatures in buildings and urban areas.

As you move to hotter drier climates, such as the middle East, the urban form become more compact, which you may expect to prohibit the ingress of wind and the transport of heat away from the urban area. However, in such areas, the urban layout has evolved so that street canyons are aligned with cooling night-time breezes to purge heat. This is possible because in continental locations with reduced topography, wind direction is generally very predictable, with winds only coming from a small range of angles. This can be seen in figure 6.24, which shows radar plots of wind direction and speed or *wind roses* for two cities in Saudi Arabia.

Aligning street canyons roughly east–west, will allow wind to flow through these street canyons allowing the urban fabric to be cooled rapidly at night. This is in contrast to the UK, which, as shown in figure 6.25, has a much less predictable wind direction. This makes it much harder for new developments in countries such as the UK, where wind speed and direction are more varied, to design urban areas to allow consistent wind ingress and to shed heat at night.

Another way in which the design of urban areas can be adapted to reduce the UHI is by maximising the sky view factor of buildings and façades. The sky view factor is a dimensionless value that ranges from 0 to 1 and is the fraction of sky

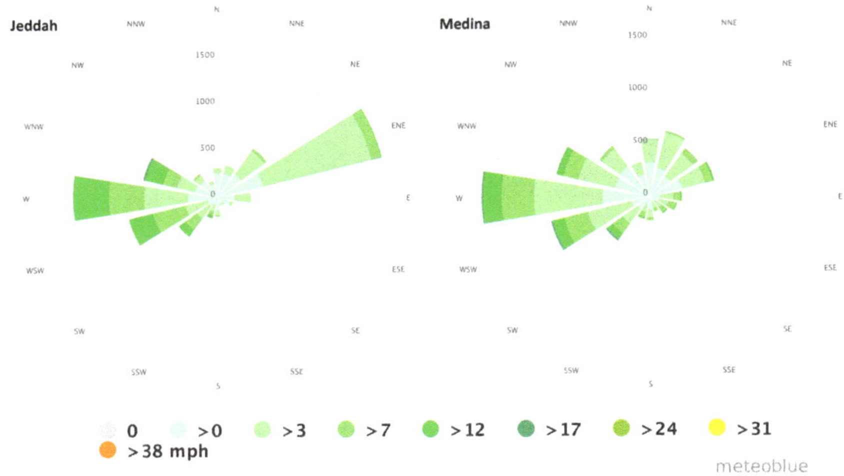

Figure 6.24. Wind roses for two cities in Saudi Arabia (Jeddah and Medina). Data from www.meteoblue.com.

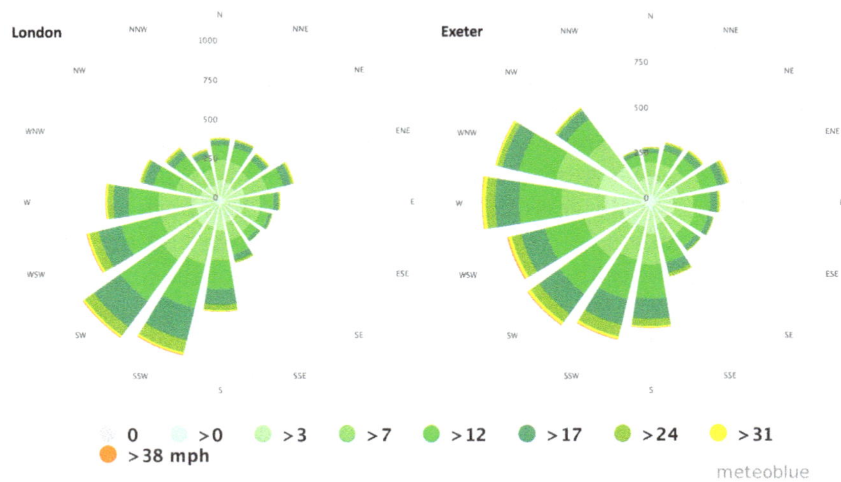

Figure 6.25. Wind roses for two cities in the UK (London and Exeter). Data from www.meteoblue.com.

visible from the ground. The denser or taller buildings are, the lower the sky view factor (figure 6.26).

If we remember back to the previous chapters, we recall that although shortwave solar radiation can be reflected by surfaces, conserving momentum, until it is absorbed or leaves the atmosphere, longwave thermal infrared radiation given off by surfaces is radiated in all directions equally. This means that for a warm surface, the sky view factor can be viewed as the probability that the radiation will leave the street canyon and pass out to space. The lower the sky view factor, the more likely that the thermal radiation will be absorbed by another urban surface, contributing to the UHI. The sky view factor of a given street canyon or building façade can

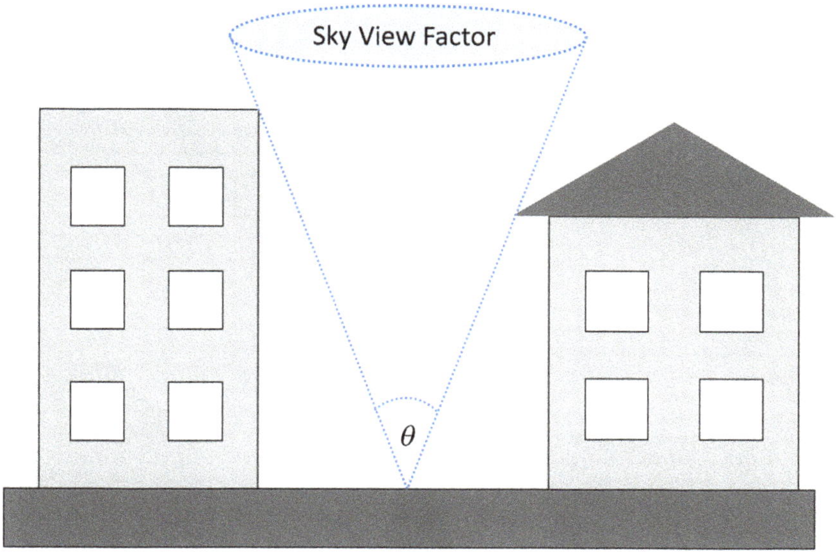

Figure 6.26. Schematic of the sky view factor principle.

easily be assessed with the use of a camera equipped with a fisheye lens, which can see all the directions in which heat could be radiated. The sky view factor is then simply the percentage of sky visible in the image, as is shown in figure 6.27.

The technique of maximising the sky view factor as part of urban planning has, for example, been implemented in Ho Chi Minh City (Vietnam). In Ho Chi Minh, green spaces are surrounded by small buildings with medium sized buildings behind and tall buildings beyond, which then decrease in size again towards the next area of green space. This effectively maximises the sky view factor for all locations and allows maximum penetration of cool air from the park into the surrounding urban form. The aim of this urban planning policy was to allow all buildings to be able to 'see the green spaces' and to be able to effectively shed heat as thermal radiation. In Ho Chi Minh, this was implemented across a large area of the city, which was undergoing rapid growth. However, this sort of intervention is not only limited to areas which are undergoing rapid urbanisation, it could be implemented in existing cities as they undergo natural demolition and redevelopment by limiting building heights. There are, however, competing issues of densification and increasing land prices to consider. Additionally, if we recall back to earlier in this chapter and the work of Ibrahim *et al* [2] on optimising urban geometry in Cairo, increasing sky view factor, while useful for shedding heat at night and reducing the nocturnal UHI, will lead to greater solar exposure during the day, particularly at lower latitudes where the solar height is greater. As such, there is a trade-off which needs to be considered —Do we reduce the nocturnal UHI? Or, do we try and limit daytime temperatures and promote outdoor thermal comfort?

Another way in which the UHI can be mitigated is through the use of urban planning to alter the surface roughness. If we recall from the previous chapter, the

Figure 6.27. Skyview image of 5th Avenue, Manhattan, New York taken with a fisheye lens. This [Flatiron fisheye] image has been obtained by the author from the Wikimedia website where it was made available by [Autopilot] under a CC BY-SA 3.0 licence. It is included on that basis. It is attributed to [Autopilot]. https://commons.wikimedia.org/wiki/File:Flatiron_fisheye.jpg

boundary layer depth δ depends upon the upwind fetch x and the surface roughness over a distance of $\sim$20–100$\times$ δ. The boundary layer is effectively determined by the average surface roughness over the fetch and that the roughness sublayer gives way to the inertial sublayer at roughly twice the average height of the roughness features (buildings or trees). It is, therefore, possible that trees or buildings that are significantly larger than their surroundings will penetrate into the inertial sublayer, creating turbulence and increasing heat loss. There could, therefore, be a benefit in dispersing tall buildings throughout urban areas to promote this effect, rather than clumping tall buildings together and enhancing the heat island. For example, the Shard in London is surrounded by smaller buildings and will likely penetrate the inertial sublayer increasing turbulence, dispersing air pollution and increasing convective heat loss. Whereas, the Eastern Cluster of the City of London on the opposite side of the river Thames, which contains several tall buildings in close proximity to each other, will enhance the UHI and lead to increased air pollution and temperatures.

The magnitude of the UHI for larger cities is equivalent in magnitude to all but the most aggressive projections of future climate change. If the UHI can be reduced through the incorporation of green and blue infrastructure or through urban planning policies, then it may be possible to offset some of the impacts of climate change. The most suitable intervention will depend upon not only the background climate of the city but also what other ecosystem services or public amenities may be of most benefit to the populous. Since climate change will alter the patterns of rainfall over the year, leading to prolonged periods of warm weather with little rainfall, and drought resistant plants have reduced evapotranspirative cooling potential, it is best to incorporate green infrastructure as part of a water retention SuDS scheme so that evapotranspirative cooling can be provided in the summer months when it is most needed. In this way, urban planning links the issues of urban water management and UHI reduction, allowing common solutions to be implemented in urban areas in the form of green and blue infrastructure.

References

[1] Oke L T R 2004 *Initial Guidance to Obtain Representative Meteorological Observations at Urban Sites* **vol 81** Geneva, World Meteorological Organization p 51

[2] Ibrahim Y, Kershaw T, Shepherd P and Coley D 2021 On the optimisation of urban form design, energy consumption and outdoor thermal comfort using a parametric workflow in a hot arid zone *Energies* **14** 4026

[3] Gunawardena K R, Wells M J and Kershaw T 2017 Utilising green and bluespace to mitigate urban heat island intensity *Sci. Total Environ.* **584** 1040–55

[4] Ampatzidis P Kershaw 2020 A review of the impact of blue space on the urban microclimate *Sci. Total Environ.* **730** 139068

[5] Ampatzidis P, Cintolesi C, Petronio A, Di Sabatino S and Kershaw T 2022 Evaporating water body effects in a simplified urban neighbourhood: a RANS analysis *J. Wind Eng. Ind. Aerodyn.* **227** 105078

[6] Ampatzidis P, Cintolesi C and Kershaw T 2023 Impact of blue space geometry on urban heat island mitigation climate *Climate* **11** 28

IOP Publishing

Climate Change Resilience in the Urban Environment (Second Edition)

Tristan Kershaw

Chapter 7

Weather extremes

Climate is a difficult thing to comprehend. We tend to think in terms of daily and seasonal variations, which we call weather. Extreme events such as heavy rainfall causing floods, snowfall causing travel disruptions or heatwaves stick in our minds. The year 2003 was not particularly warm. In fact, the summertime temperatures were about average. However, the two-week heatwave that sat over Europe in August 2003, which claimed ~70 000 lives, biases our opinion about the summer of 2003. Such events are extreme not only in that they occur infrequently but also that they have profound impact on people and urban areas. As the climate changes as a result of human activity and anthropogenic carbon emissions, the relative probabilities and likelihoods of extreme events will change. The intensity of such events will also change, with hotter, longer heatwaves and more powerful storms, with higher wind speeds and more rainfall, so that what is currently classed as extreme weather may become the norm in the future. Unfortunately, the low probability nature of extreme weather means that there are few events in the historical records, which makes predicting future reoccurrences difficult. As such, there is a lot of uncertainty in climate change projections when it comes to lower probability events such as powerful storms and heatwaves. However, due to their destructive nature to property, infrastructure and human lives, there is considerably ongoing research to improve the future forecasting of such events.

7.1 Heatwaves

Heatwaves are periods of calm hot weather (for that location) that persist for several days, which may also be accompanied by high humidity in marine climates. There are several different definitions of what constitutes a heatwave depending upon the region. A common definition that has been used for the UK is three or more consecutive days with daily maximum temperatures above 28 °C, not dropping

doi:10.1088/978-0-7503-5262-8ch7

below 16 °C at night. This definition is useful because it highlights the risk to human health in that the human body is able to withstand high temperatures for short periods, so long as it is able to shed stored heat when temperatures drop, usually at night. Another common and more international definition of a heatwave is the one used by the World Meteorological Organisation (WMO), which is five or more consecutive days where the daily maximum temperature of exceeds the average maximum temperature by 5 °C, the baseline period being 1961–90. This definition is useful because it recognises that heatwaves may not always be in the height of summer and that a particularly hot spring or autumn can be just as dangerous as higher temperatures in summer, due to the time lag associated with the adaptive response of human physiology. However, it does ignore the dangers associated with higher daily minimum temperatures and the associated implications for human health and wellbeing.

Estimates of the number of deaths across Europe as a result for the 2003 heatwave vary from ~15 000 up to ~70 000 depending upon methodology used and the assumptions made regarding how many people would have died anyway during that period or in the following few months. Regardless of the methodology used to estimate the death toll, heatwaves have a profound impact on the populous, especially on vulnerable groups such as the elderly, infirm or very young. The deaths that occurred as a result of the 2003 heatwave, mainly occurred in large urban areas with a distinct UHI, which itself is intensified by the atmospheric conditions typical of a heatwave. The August 2003 heatwave was thought to be the warmest period for up to 500 years and many European countries experienced their highest temperatures on record. In the UK, temperatures peaked at 38.5 °C and the London UHI was measured to be 9 °C. According to the UK Office for National Statistics, there were 2091 excess deaths due to the 2003 heatwave across the UK. As we have discussed in previous chapters, high temperatures and reduced thermal comfort have a negative impact on worker productivity. As a result, the 2003 heatwave is estimated to have cost the UK economy between £400 m and £500 m in reduced manufacturing output.

Following the 2003 heatwave, the UK put into place a heatwave plan and several procedures to alert the public of the danger of extreme heat with a colour coded warning system (Green, Yellow, Amber and Red). Under the new criteria, several periods over the last few years have been classed as 'Heat Periods' (see table 7.1). Heat periods use the UK Health Security Agency (UKHSA) definition of day(s) on which there is an Amber (level 3) heat health alert, or day(s) upon which the average Central England Temperature (CET) is above 20 °C. This is different from the definition of heatwaves used by the WMO and others, and reflects the level of risk and the potential impact of high temperatures, instead of just a threshold temperature. As such, only a few of the recorded heat periods will be classed as heatwaves under the WMO criteria. Recently, the 2022 heat periods, of which there were five periods of extreme heat over the summer, saw the highest temperature record in the UK broken on the 19 July 2022 with 40.3 °C recorded in Coningsby, Lincolnshire, beating the previous UK peak temperature record by 1.6 °C. New regional records were set across all of England, Wales and Northern Ireland, with many places

Table 7.1. Summary statistics of excess mortality during heat periods using UKHSA definition, 2016–22. Crown Copyright. Contains public sector information licensed under the Open Government Licence v1.0

Heat period dates	Excess deaths	Percentage excess deaths (%)	Excess deaths (< 70 years)	Percentage excess deaths (%) (< 70 years)	Excess deaths (> 70 years)	Percentage excess deaths (%) (> 70 years)
23–25 August 2022	−520	−13.0	−275	−29.5	−245	−8.0
8–17 August 2022	1279	9.7	−413	−13.4	1693	16.9
30 July–5 August 2022	10	0.1	−396	−18.3	403	5.8
10–25 July 2022	2227	10.4	−525	−10.5	2750	16.8
16–19 June 2022	275	5.1	−140	−11.5	416	10.1
6–9 September 2021	1453	27.7	311	25.9	1141	28.3
16–23 July 2021	2030	19.5	461	19.0	1568	19.6
5–15 August 2020	1347	9.5	222	6.6	1126	10.4
30 July–1 August 2020	319	8.5	53	5.9	267	9.4
23–27 June 2020	642	9.9	94	6.0	550	11.1
23–29 August 2019	582	6.5	86	4.0	496	7.2
21–28 July 2019	692	6.8	−1	0.0	694	8.9
28–30 June 2019	161	4.2	−9	−1.0	170	5.9
2–9 August 2018	306	3.0	22	0.9	284	3.7
21–29 July 2018	738	6.6	−7	−0.3	747	8.8
30 June–10 July 2018	557	4.0	101	3.0	453	4.2
25–27 June 2018	386	10.3	36	3.9	350	12.3
5–7 July 2017	415	11.0	72	7.9	343	12.0
17–23 June 2017	1112	12.5	149	7.0	962	14.2
12–17 September 2016	268	3.5	103	5.6	167	2.9
22–26 August 2016	631	10.3	153	10.4	479	10.2
18–22 July 2016	959	15.4	108	7.2	852	18.1

exceeding 40 °C for the first time. This same heatwave saw temperatures of over 45 °C in Spain and 47 °C in Portugal.

During the five heat periods between June and August shown in table 7.1 (referred to as H1 to H5) there were 56 303 deaths across England and Wales. This is 3271 deaths above the five-year average for this time of year, and considerably higher than the 2003 total of 2091 excess deaths. This is an average of 82 excess deaths per day, which is 6.2% higher than the five-year average. The heat period with the greatest number of deaths was H2, which covered the period 10–25 July 2022 with 2227 excess deaths, an average of 139 excess deaths per day, which is 10.4% above the five-year average. H2 was characterised by a brief cooling (albeit not below the threshold temperature for a heatwave) and then a sharp increase in temperature peaking on the 19 July. The 2022 data should be treated with a certain amount of caution due to the Covid 19 pandemic. Deaths of persons who had registered a positive Covid test within the previous 28 days were discounted. Furthermore, the five-year averages used to calculate the excess deaths will have been affected by the pandemic. It is also possible that data for heat periods during the pandemic will be artificially inflated. This creates a certain amount of uncertainty in the exact figures of excess deaths reported by the Office for National Statistics post-pandemic; however, examination of the reported figures is still interesting. As such we will confine ourselves here to pre- and post-pandemic.

Heatwaves by definition are not a common occurrence, but they are becoming more common due to a warming and more variable climate. However, many heatwaves are not as disastrous as the 2003 heatwave. This is likely due to the preceding 2003 summer months being abnormally cool, with sudden changes in temperature being more dangerous to humans. We can see from Table 7.1 that there is considerable variation in the number of excess deaths recorded during the heat periods between 2016 and 2022. We can see that the number of deaths of people aged over 70 years is nearly always higher than those under 70. This is an indication of the vulnerability of the more aged population. We can also see where heat periods are close together there is a harvesting effect, where we have instances of negative excess deaths, where the number of deaths recorded is lower than the five-year average. This is an indication that the prior heat period has already ‘harvested’ individuals who are susceptible to heat-related illnesses and there is currently a deficit. This is most apparent in the H5 event of 2022 (23–25 August), where the population both the over and under 70 years show negative excess deaths, indicating that the prior four heat events had already harvested a large proportion of heat-vulnerable individuals. We can also see that the under 70 years values for 2022 are all negative, which may well be an artefact from the pandemic skewing the number of deaths in younger people in the five-year average. What is apparent form the data in table 7.1 is that older populations are more susceptible, and that the magnitude of the heat event is directly related to the number of excess deaths.

Global populations are aging with people living longer in both developed and developing countries, and the UK is no different. The percentage of the UK population aged 75 or over is expected to increase from ~8% in 2015 to ~18% by 2085. When temperatures exceed 35 °C, all age groups are at risk from the high temperatures; however, it is the elderly (and other vulnerable groups) who are most at risk and likely to be adversely affected by extremes of temperature. It is entirely likely that humans will undergo some form of autonomous physiological acclimatisation to changes in mean temperatures over time. However, research suggests that humans are less able to adapt to rapid, sudden changes in temperatures, particularly if the overall annual temperature variability increases as it is expected to as a result of climate change. If we examine table 7.1, then we can see that there are periods of high mortality. June 2017 was officially classed as a heatwave under the WMO criteria, as were periods of 2018, although continental Europe bore most of the brunt of this particular heatwave. Both indicate an elevated number of deaths compared to other heat periods. This is of concern because both the frequency and intensity of heatwaves are expected to increase as a result of climate change. Current estimates indicate that the number of heat-related deaths in the UK over a typical year could increase from the ~2000 now to ~7000 by 2050.

Cold is still the main cause of temperature related deaths in the UK, with between 35 800 and 49 700 cold-related deaths per year on average in the UK. As the climate warms, cold will remain an important climate risk, even though winters will become milder overall because minimum temperatures will remain similar to those we experience now. This coupled with the generally poor thermal performance of the UK housing stock and a growing, ageing population means that the number of cold-related deaths is unlikely to change much in the near future. Current estimates indicate only a ~2% reduction in cold-related deaths in the UK by 2050 under a medium emissions scenario (A1B). Current measures to improve the energy efficiency of homes will also reduce the impact of cold weather on the occupants. However, increasing the airtightness of the building fabric and increasing the level of insulation has to be undertaken carefully so as not to increase the risk of overheating, provision needs to be made to allow the removal of heat in the summer months.

The number of hot-days per year has been increasing steadily since the 1960s and currently there are around 2000 heat-related deaths in the UK each year. As the number of hot-days per year increases as a result of climate change, so does the risk of a severe heatwave. All but the lowest estimates of climate change (RCP2.6) show that by the 2040s a summer as hot as 2003 will be very common and will be considered cold by the end of the century [1]. The ageing population means that the number of vulnerable people is increasing, which means that the number of premature heat-related deaths is also expected to increase. The number of heat-related deaths in the UK is predicted to more than triple by the 2050s and increase five-fold by the 2080s under a medium scenario (A1B, median estimate) from the baseline of 2000 deaths per year. This could be further exacerbated by poorly implemented energy saving measures such as increased airtightness and insulation unless also accompanied by increased ventilation provision (for example). There is evidence that newer homes are at a greater risk of overheating than older designs, due to not only improved airtightness and greater insulation but also due to the removal or masking of thermal mass by use of modern construction materials and methods. At present, there are no comprehensive policies in place to adapt the existing building stock to higher temperatures, reduce the urban heat island or to safeguard homes against heatwaves. There is evidence that most people lack a basic understanding of the risks to health from elevated indoor temperatures, and therefore are unlikely to take measures to safeguard their and their dependents' wellbeing. Particular care will therefore need to be applied to buildings such as schools, hospitals, care homes and prisons.

Ongoing research into the return periods of extreme weather events such as heatwaves has shown that human activities have increased the likelihood of heatwaves considerably, an event that would have been expected to occur twice a century in the early-2000s can now be expected to twice a decade. A statistical analysis by Christidis *et al* [1] showed that the effects of climate change increased the likelihood of a more severe 2003 type event from 1 in several thousand to ~1:100 in little over a decade and by 2040 this can be expected to be an average summer. As the climate warms, the distribution of the summer temperatures is expected to not only shift to higher warmer regimes but also stretch, increasing overall variability in temperature, which would further increase the frequency of heatwave events (using the current definitions). Estimates of the effects of climate change on heatwave events are varied. For a 2 °C increase in global mean temperatures, it is estimated that a heatwave event in Europe may be expected to increase in intensity by between 1.4 °C and 7.5 °C [1].

The situation though is more complex than just increasing heatwave intensity. Heatwaves are typified by low wind speeds, intense solar radiation and atmospheric stability, as well as high temperatures. This means that the UHI is exacerbated by heatwaves, further worsening conditions in urban areas. In December 2023, Australia was experiencing a heatwave and cities such as Sydney were also experiencing UHIs of ~10 °C. Western Sydney is one of the country's fastest growing areas and also has one of the fastest growing poverty rates. This is where the impact of extreme weather becomes doubly problematic because its effects are not felt equally. Underprivileged communities cannot necessarily afford air conditioning, buildings may well be constructed to a lesser standard and be more vulnerable to higher temperatures, with less green or blue space help cool the air. The elderly, very young and those with underlying medical issues are typically at greatest risk from high temperatures and it is generally accepted that more deprived communities have worse health outcomes than those who live in more affluent areas. In the case of Western Sydney, where 2.5 million people live, the Blue Mountains shield the area from cooling coastal breezes which could alleviate the accumulated heat, many residents live in buildings with poor levels of insulation (here to keep heat out rather than keep heat in), buildings have dark coloured roofs and there is little in the way of vegetation. The lower levels of income per capita in the region means that many cannot afford air conditioning. This is not an easy issue to fix either because improving the building stock or adding vegetation and green/blue infrastructure to alleviate peak temperatures will increase the desirability of the area, increase land prices, and ultimately force out the most vulnerable in the community. This is a process known as gentrification and is a major factor when considering climate justice in an urban environment.

The methods we can employ to keep our buildings cool and safeguard ourselves against high temperatures are similar to those already covered in chapters 4 and 6. Due to the characteristics of a heatwave (high temperatures, intense solar radiation and high atmospheric stability with little wind), ventilation may be inefficient or even harmful unless the air can be cooled somehow (e.g. by evapotranspiration from plants) because the external air may be hotter than that inside the building. Due to reduced wind speeds, cross ventilation may well be ineffective at removing heat. Furthermore, thermal mass, which is effective at reducing peak temperatures, may well contribute to high mean internal temperatures during a prolonged heatwave, particularly if exposed to direct sunlight, unless stored heat can be purged at night. Effort is best placed at keeping heat out of our buildings, closing curtains blinds or shutters early in the day for instance to keep solar radiation out of the building, particularly west facing rooms which will get afternoon and early evening sun, because these will be the most prone to overheating. External shading will also be effective, whether in the form of roof overhangs, balconies, brise soleil or even just trees and vegetation will reduce the ingress of solar radiation and reduce internal temperatures. During a heatwave where nocturnal temperatures do not drop significantly at night, building heat loss via radiation and convection will be reduced at night, as such windows should be left open to allow heat to escape from the structure, purging any thermal mass, this will be enhanced due to the pressure difference between warmer air inside the building compared to outside. If possible, openings on multiple storeys should be used to also utilise the change in pressure with altitude (so-called stack ventilation); however, ground floor openings can be a security risk. It is imperative that these windows or openings are closed again in the morning before external temperatures increase to retain the coolth within the building. Changing the colour of the building will help reduce the risk of overheating by reflecting solar radiation. Additionally, increasing the amount of insulation in your loft will assist in keeping homes cool because the majority of solar heat gains are via the roof space.

Finally, more drastic measures to reduce overheating may focus on adding or increasing access to thermal mass within the building structure (coupled with effective night-time ventilation). Modern building materials such as plasterboard are effective insulators and a dry-lined wall effectively masks any thermal mass behind (the same is true of suspended ceiling tiles). Replacing plasterboard with wet gypsum plaster allows heat to be absorbed from the air by the bricks/blockwork behind. Similarly replacing carpets or wooden flooring with stone flagstones or ceramic tiles will allow heat to be stored in the screed/concrete floor. These effectively reduce not only the air temperature but also the mean radiant temperature because the surfaces around us are at a lower temperature. This has the effect of making us feel cooler because our bodies, which are almost perfect receivers/absorbers of thermal radiation (a black-body), are receiving less radiation from our surroundings, and hence are able to radiate heat more effectively. Ultimately, any heat stored within the building structure needs to be lost to the environment eventually. Either the building needs to be made so thermally massive that the building will take several weeks to reach equilibrium with the elevated external temperatures (e.g. complete stone or concrete construction such as a church) or else ventilation needs to be deployed to purge heat at night when temperatures are lower. Increasing the ventilation provision of a building by adding secure floor and ceiling level vents (preferably on opposite façades) will allow stack driven ventilation to occur at night, removing heat from the building and purging thermal mass ready for a new day.

7.2 Storms

When I initially wrote this section in 2017, storm Aileen was battering my house with 100+ $km\,h^{-1}$ winds and torrential rain. While simultaneously across the Atlantic hurricane Irma was causing widespread devastation to Florida after already passing through the Caribbean and the Bahamas. In 2024, as I write the second edition of this book, storm Jocelyn is raging outside. The severe impacts of high winds, torrential rain and storm surges seem highly evident. The number of severe storms per decade has been rising steadily since the 1950s, with the stormiest years coinciding with a sustained positive North Atlantic Oscillation (NAO) index, a measure of sea-level pressure anomaly versus the long-term average. A positive value of NAO indicates higher pressure and is typical of stronger westerlies over the mid-latitudes, more intense weather systems and wetter and milder weather over western Europe. Whereas a negative NAO blocks storms moving up from the Gulf of Mexico towards the UK and the least stormy years are associated with more negative NAO pressures. It seems evident that there is an association between the NAO and the El Niño/La Niña Southern temperature oscillations over the Pacific, but this relationship is not well understood due to the many other factors involved, such as sea ice melt in the Arctic for instance. Unfortunately, this makes predicting future storms and wind speeds very difficult, particularly when considered alongside the other uncertainties associated with estimates of climate change. This means that this section is based upon less evidence than the previous sections of this book; however, the risks to buildings, infrastructure and human life form powerful storms require that it is given consideration. Furthermore, it seems logical that, given the trends of increased winter rainfall and increased temperatures, the number and intensity of storms will increase.

People still talk of the 'Great Storm of 1987', which devastated the south of England. Schools were evacuated and pupils sent home (myself included), and people were advised to stay at home where it was safe, while outside trees were uprooted, fences and walls were blown over and roofs were ripped off buildings. In the coastal village where I lived, sea defences were destroyed and sections of coastal roads and footpaths washed away. This storm is considered to herald the onset of increased storminess, with the 1990s being the UK's stormiest decade since the 1920s.

The UK experienced a spell of extreme weather during the winter of 2013/14. Between late-January and mid-February 2014, a succession of major storm brought flooding and destruction to the UK. During these few weeks around six major storms separated by two to three days hit the UK, these followed an earlier stormy period during December and early January. Overall, during this winter period

(December–February) at least 12 major winter storms hit the UK, making this the stormiest period for at least 20 years. January 2014 was the wettest January on record with more than 5 months' worth of rainfall recorded in some locations. The most severe of this rapid succession of storms occurred in early- to mid-February with ground still saturated and inundated from rainfall over the new-year period.

These storms resulted in numerous weather-related impacts across most of the UK throughout the winter period. There was major flooding, with the Environment Agency reporting 6000+ properties flooded. The Somerset Levels were very badly affected, with large areas remaining underwater from late-December through the entire winter period to mid-March. There was also severe flooding along much of the River Thames through Oxfordshire, Berkshire and Surrey. In addition to flooded to properties and businesses, transport infrastructure was also affected with many roads underwater and several villages on the Somerset Levels being only accessible by boat. The Southwest mainline railway between Exeter and Bristol was also affected by flooding on the Somerset Levels, being shut until mid-March. The following video clip from the Met Office shows the extent of some of the impacts that these storms caused (figure 7.1).

As well as inland flooding, strong winds and huge waves battered exposed coastlines in the South and Southwest causing widespread destruction to property and infrastructure. Strong winds, high tides and tidal surges acted in combination to produce huge waves to batter the coastline (see figure 7.2). The swells produced by the storms had a long wavelength, allowing for the formation of waves with a large amount of speed and energy, and able to reach record heights. The most severe storm during this winter period was on the 12 February, with waves of 25 m (82′) recorded off Southern Ireland and wind speeds reaching hurricane force (see table 7.2).

Figure 7.1. Weather summary video clip of winter 2013/14 (Crown Copyright. Contains public sector information licensed under the Open Government Licence v1.0). Video available at http://iopscience.iop.org/book/mono/978-0-7503-5262-8.

Figure 7.2. Huge waves batter the Cornish village of Porthleven on 5 February 2014. ©Matt Clark, Met Office

As we can see from figure 7.2, coastal conditions can become exceptionally dangerous. This is not only a danger to human life but can also cause considerable damage to buildings and infrastructure as huge waves overtop coastal defences. The mainline railway through Devon to Cornwall was severely damaged along a coastal stretch of line at Dawlish during the storm of 4–5 February (see figures 7.3 and 7.4), severing a key transport link to the Southwest peninsula for several weeks. This sort of disruption has a knock-on effect for the whole of the rail transport network because many national rail services begin or end in the Southwest along the stretch of line beyond Dawlish. This meant that a significant proportion of the rolling stock became trapped on the Southwest peninsula, causing widespread disruption across the entire rail network. These storms caused considerable coastal erosion and the rapid succession of storms exacerbated coastal damage as weakened defences could not be repaired before the next storm hit.

Figure 7.4 shows damage to the substantial retaining wall supporting the foundations to the mainline railway. This damage caused by waves overtopping coastal defences is likely to be a result of the succession of storms progressively weakening the wall. This highlights the need for a more rapid response after storms to check and repair sea defences and infrastructure. It would seem probable that as the climate warms and contains more energy, the weather will become turbulent, resulting in increasingly powerful storms. A lack of preparedness will only exacerbate the effects of extreme weather.

The winter of 2015/16 was also exceptional. The mildest December on record was also the wettest, this was coupled with the seemingly continuous series of storms that swept across the UK. In total, 509.2 mm (~20 inches) of rain fell in the three-month period December 2015 to February 2016. December 2015 was the warmest December since records began in 1910 with mean temperatures about 4 °C above the long-term average. The UK Met Office state that there is a direct link between the warmth and the record levels rainfall, with December being also the wettest of

Table 7.2. The Beaufort scale used to describe wind intensity.

Beaufort wind scale	Wind description	Wind speeds in mph (km h^{-1})	Sea description	Probable wave height (max probable) in m
0	Calm	<1 (<1)	Calm (glassy)	0 (0)
1	Light air	1–3 (1–5)	Calm (rippled)	0.1 (0.1)
2	Light breeze	4–7 (6–11)	Smooth (wavelets)	0.2 (0.3)
3	Gentle breeze	8–12 (12–19)	Slight	0.6 (1)
4	Moderate breeze	13–18 (20–28)	Slight–moderate	1 (1.5)
5	Fresh breeze	19–24 (29–38)	Moderate	2 (2.5)
6	Strong breeze	25–31 (39–49)	Rough	3 (4)
7	Near gale	32–38 (50–61)	Rough–very rough	4 (5.5)
8	Gale	39–46 (62–74)	Very rough–high	5.5 (7.5)
9	Strong gale	47–54 (75–88)	High	7 (10)
10	Storm	55–63 (89–102)	Very high	9 (12.5)
11	Violent storm	64–72 (103–117)	Very high	11.5 (16)
12	Hurricane	73+ (118+)	Phenomenal	14+ (–)

Figure 7.3. Damage to the Southwest mainline railway at Dawlish (Devon) after the storm of 4–5 February. Breach of sea defences and destruction of retaining wall, left rails suspended in mid-air after foundations were washed away. ©Matt Clark, Met Office

Figure 7.4. Photo showing damage to the retaining wall bounding the mainline railway at Dawlish. ©Matt Clark, Met Office

any calendar month of record with almost double the normal amount of rainfall at 218.8 mm.

In 2015, the UK Met Office began naming storms rather than just hurricanes, with the first named storm being Abigail, which hit UK shores on the 10 November

2015. Over the next few months there was a rapid succession of storms (see table 7.3), which caused disruption and flooding. The three storms in December provided much of the total month's rainfall, with the storms propelled by the jet stream towards the UK, with additional contributions from the El Niño weather phenomenon and anthropogenic climate change. The jet stream which flows from the Gulf of Mexico towards the UK is responsible for much of our mild weather and is indicative of a positive NOA. In this case, the positive NOA and strong jet stream will have contributed towards the uncommonly mild weather, with the elevated temperatures resulting in greater evaporation of water, leading to increased rainfall.

Subsequent storm seasons have seen variation in the number of storms hitting UK shores, ranging from four in 2022/23 to 10 storms in 2017/18. The 2023/24 storm season has already proven interesting with 10 storms to date, with storm Jocelyn currently raging outside (24 January 2024). Storm Ciarán (1–2 November 2023) saw a Red weather warming (defined as 'danger to life') issued by the Jersey Met Office, and an Amber warning across most of the southern UK, resulting in the closure of hundreds of schools. This storm made headlines with it being compared to the 'Great Storm' of 1987, and was described as a 'weather bomb' due to the explosive power caused by a rapid drop of pressure, a process called explosive cyclogenesis, whereby central air pressure drops by more than 24 millibars in 24-h. Storm Ciarán was expected to see a drop of 28 millibars, which prompted the flurry of weather warnings. While this storm generated high wind speeds, with gusts of 119 mph (193 $\mathrm{km\,h^{-1}}$) in Brittany, France, the high winds were generally confined to offshore areas, with maximum wind speeds of 70–80 mph (112–129 $\mathrm{km\,h^{-1}}$) recorded on land which are typical of a major Atlantic storm. By comparison, the 'Great Storm' saw on shore gusts of 98 mph (158 $\mathrm{km\,h^{-1}}$). While storm Ciarán did not quite deliver on the hype generated by the weather warmings, it did set a new record for the lowest pressure of 953.3 millibar at Plymouth (Devon). Such a low pressure is exceptionally unusual for southern England. The previous record of 959.7 millibar recorded at

Table 7.3. Names and dates of the 11 storms in the 2015/16 storm season, currently the greatest number of storms in one season since naming started.

Name	Date of impact on UK and/or Ireland
Abigail	12–13 November 2015
Barney	17–18 November 2015
Clodagh (Clo-da)	29 November 2015
Desmond	5–6 December 2015
Eva	24 December 2015
Frank	29–30 December 2015
Gertrude	29 January 2016
Henry	1–2 February 2016
Imogen	8 February 2016
Jake	2 March 2016
Katie	27–28 March 2016

Teignmouth (Devon) had stood since 1916. The Met Office recorded a sharp drop in pressure from >990 millibar down to 953.3 millibar in 12-h before rising again. Such a sharp drop in sea-level pressure is indicative of powerful storm systems, which generate high winds and driving rain. With warming oceans and more energy present in the climate system we can expect storm events to become increasingly powerful, if not more frequent also. We will no doubt see more storms compared to the 'Great Storm' of 1987 in the near future.

Building codes in the UK already take account of the impact of wind shear forces on buildings. However, projections of climate change indicate that storms will become more powerful, meaning that additional care will need to be applied to the design of buildings in exposed areas where squalls of very strong winds are common, even if they are typically only of a short duration. Violent storms have a variety of names according to the region in which they occur; hurricanes, typhoons and cyclones are all basically the same phenomenon, consisting of high speed ~120 km h^{-1} (~75 mph) rotating winds. This is mean wind speed; individual gusts can be significantly stronger and more powerful. These can cause extensive damage (table 7.4) that can be caused not only by the direct effects of the powerful winds and intense, torrential rainfall, but also indirectly by associated flooding, storm surge and landslides.

The energy contained within wind, and hence its destructive power, increases with the square of its velocity; hence, a 100 km h^{-1} gust is 100× greater force than a 10 km h^{-1} gust. To resist the lateral forces of these winds, buildings are typically made more rigid or braced somehow. However, more primitive architectures in tropical climates adopt the philosophy of flexibility, allowing structures to sway and

Table 7.4. Wind gust velocity and damage to vegetation and property.

Weather extremes	Weather extremes
75 mph–93 mph 120 km h^{-1}–150 km h^{-1}	Shrubs, trees and foliage but unlikely for buildings in good repair.
94 mph–112 mph 151 km h^{-1}–180 km h^{-1}	Considerable damage to shrubs and tree foliage with some trees blown down. Some damage to building roofs, windows and doors, minimal damage to building structures.
112 mph–130 mph 181 km h^{-1}–210 km h^{-1}	Extensive damage to trees and shrubs with large trees blown down. Some damage to building roofs, windows and doors, minor damage to curtain walls and small structures.
131 mph–149 mph 211 km h^{-1}–240 km h^{-1}	Shrubs and trees blown down. Extensive damage to roofs, windows, doors and some curtain walls. Complete failure of roofing structures on small buildings.
150 mph + 241 km h^{-1}+	Shrubs and trees blown down. Considerable damage to buildings; roofs, windows and doors. Complete failure of roofing structures on small buildings. Curtainwall and façade failure of industrial type buildings. Small buildings may be overturned completely.

give in the winds like the palm trees of such climates. The alternative to these solutions is to shelter then building, either via windbreaks or by reducing the profile of the building, making it more aerodynamic or sinking it in to the ground.

A building in a location prone to cyclonic storms must be able to withstand the extreme winds that can blow from every direction as the storm moves overhead. The building must also be able to withstand the rapid and extreme changes in air pressure generated by the storm. Severe storms are characterised by this rapid change in pressure, differing them from severe gales. This change in pressure is $>\pm 10$ kPa ($>\pm 10$ mb) in a 3-h period. This coupled with the local changes in pressure due to wind flow and the Bernoulli principle (cf. the lift produced by plane wings due to the pressure difference between the top and bottom) means that the pressure across different building elements can be extreme. Additionally, as we have previously discussed, the drop in pressure due to an intense cyclonic storm leads to a rising of sea levels, which can lead to coastal or estuary flooding, particularly if the storm coincides with an astronomical high tide.

Roofs are particularly prone to being damaged or destroyed by the high suction (negative pressure difference) on their leeward side (see figures 7.5 and 7.6) due to the rapid flow of air over the building or a combination of suction of top and pressure underneath the roof overhangs. In taller buildings, the influence of wind gusts and the pressure differences between the windward and leeward sides of a building can set up oscillations in the building structure, which can result in structural failure.

The centre of a cyclonic storm is at extremely low pressure and if this hits a building that is particularly well sealed too quickly, the normal pressure (or even elevated pressure due to temperature differences) inside the building can cause it to explode outward and collapse. This is perhaps an extreme example but the failure of window and door seals (for example) leading to increased air leakage (infiltration) and potentially increased energy bills in the longer term if not remedied is much more likely. This presents a problem because the current tendency is to create more insulated and airtight buildings on the pretext that they are more energy efficient.

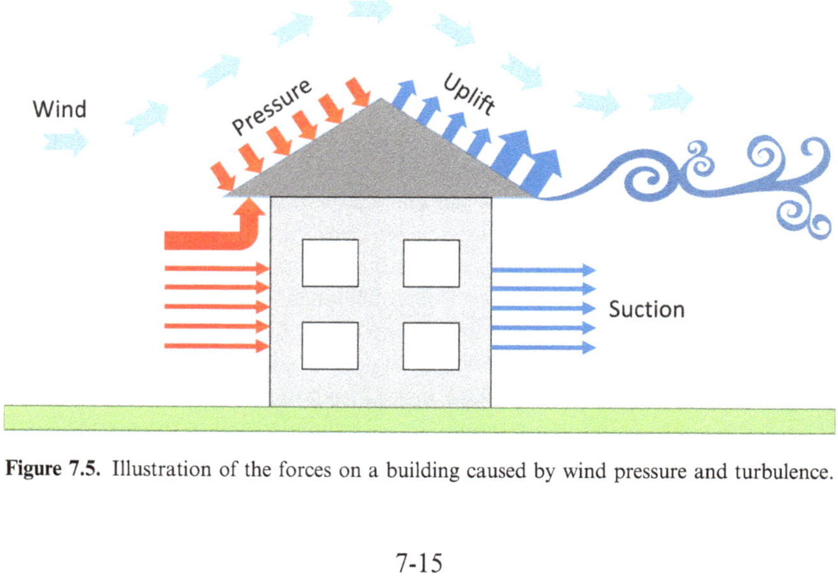

Figure 7.5. Illustration of the forces on a building caused by wind pressure and turbulence.

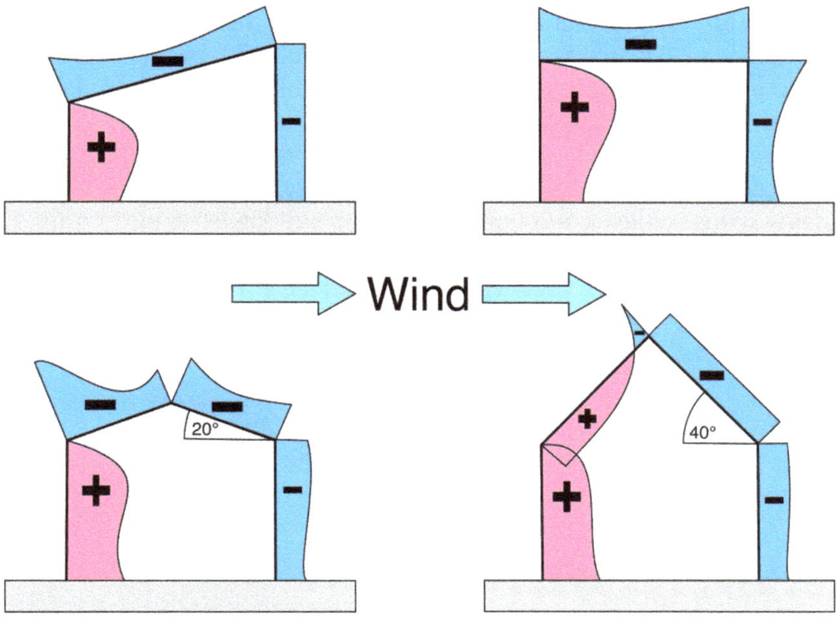

Figure 7.6. Forces exerted on different roofs and façades by the wind, positive indicates pressure while negative indicates suction.

Buildings can be protected from high winds by planting trees, but care should be taken not to plant them closer than ~15 m in case the tree is blown over. The sheltering effect of such vegetation will be effective only over a distance of ~8–10 times the height of the tree. In undulating or hilly areas, building construction along ridges should be avoided because they will experience accentuated wind velocities, whereas buildings lower down in a valley will typically experience lower speeds. The reverse, however, may be true in long narrow valleys where the wind becomes funnelled along the valley floor, this obviously depends on the orientation of the valley to the prevailing winds.

The shape of a building is the single most important factor in determining the performance of buildings in cyclonic storms. Compact, simple geometric shapes are best. Buildings should ideally be symmetrical, with square, or even hexagonal floor plans and roofs should have multiple slopes that present less area to winds and will perform better under wind load forcing than gable roofs with two slopes. If the site dictates a rectangular shape, then it is best if the length is not more than three times the width. Experiments have shown that hip roofs with a pitch in the range 25°–40° are most resistant to high winds. While roofs with a pitch <22° should be avoided due to the increased suction from wind flow (figure 7.6).

All roof elements, such as rafters and purlins, should be securely fixed to the main structure. Additionally, the double nailing of roof tiles is advisable to prevent removal by suction forces. Roof canopies and overhangs should be minimised or avoided if possible, although herein lies a contradiction with reducing temperatures by shading of façades. In some climates, brackets fixed to the main building structure or to the ground can be used to tie down the roof before the storm season starts. Verandas or covered walkways should be built as separate structures rather than extensions to the main structure so they can blow off without damaging the main structure. Windows and doors are a particular danger in high winds. If they are broken by flying debris or breached by changing pressure levels, then the wind will have easy access to the building's interior and the resultant uplift pressure in the roof space may well cause the destruction of the entire building structure. Frames should be securely anchored to the rest of the building structure and large expanses of glass should utilise shatterproof, toughened glass and be protected by storm shutters. Many of these design features can be seen employed in the traditional buildings of the world's exposed locations.

Reference

[1] Christidis N, Jones G S and Stott P A 2014 Dramatically increasing chance of extremely hot summers since the 2003 European heatwave *Nat. Climate Change* **5** 46–50

IOP Publishing

Climate Change Resilience in the Urban Environment (Second Edition)

Tristan Kershaw

Chapter 8

Conclusions

In this book, we have seen that the global climate is changing, with anthropogenic greenhouse gas emissions the dominant cause. The global increase in temperature of ~1 °C since 1880 (see figure 1.15) is also mirrored in the UK climate, with higher average temperatures (figure 1.13). Average temperatures across the UK have increased in line with global projections of climate change, with a trend towards milder, wetter winters and warmer, drier summers. Sea levels globally and around the UK have risen by between 15 and 20 cm since 1900 and there is a growing amount of evidence that climate change is partly responsible for the severity and frequency of recent extreme weather events.

With very few exceptions, the building industry has yet to face up to the implications of climate change and the effect this will have on the performance of our buildings, not just in terms of energy usage but also in terms of thermal comfort and the effect this has on the productivity, health and well-being of the building occupants. The building industry is still coming to terms with how to reduce the impact of our buildings on the climate via reducing carbon emissions and improving efficiency as part of the mitigation agenda, let alone started to consider how to reduce the long-term impact of the climate on our buildings. The buildings we design now will still be in use when the impacts of climate change make themselves felt and in the case of residential buildings many of the homes we will be occupying at the end of this century have already been built and in a lot of cases were built before the introduction of building regulations (1965). One important realisation is the fact that the weather data we base the design of our building on is out of date, being based upon the averages of historical observations, primarily from the previous century. Despite the recent efforts of academics and professional bodies such as the Chartered Institution of Building Services Engineers (CIBSE) to update this weather data, such historical observations will inevitably lag behind the changing climate.

doi:10.1088/978-0-7503-5262-8ch8

Therefore, it is becoming increasingly important to assess how buildings will perform in the future, not just based on the current or historic climate.

The Paris Agreement (2015) represents a significant step forward to curb carbon emissions and limit future climate change, 195 nations including the UK will 'pursue efforts' to limit the global temperature increase to <1.5 °C. For this ambitious goal to be achieved, global carbon emissions will need to peak soon and decline rapidly afterwards. Currently, the commitments of the partner countries will still lead an estimated 2.7 °C rise in global temperatures, indicating how much further there is still to go. The uncertainties associated with climate change predictions (see chapter 1) mean that even if the Paris Agreement targets are met, there is still the possibility of a ~4 °C warming of global temperatures by the end of the century. Furthermore, we need to consider that these are global averages and that by 2100 there will still be significant lag associated with the thermal inertia of the oceans, so land surface temperature increases will be significantly higher than 1.5 °C. We therefore need to consider how we will cope with increasing temperatures of several °C and the implications this will have on the weather and performance of our buildings, without compromising the ongoing mitigation agenda. The knowledge contained within this book is not new, much of it has been employed in various locations around the world for millennia. To increase the resilience of our buildings and urban areas, we simply need to apply it.

8.1 Building resilience

This book has considered how buildings in different locations have evolved to become resilient to the local climate, whether that climate is warm and humid, hot and dry, temperate, marine or arid. By examining the responses of different locations to the environmental stimuli, we can consider how to adapt our building design to cope with anticipated future climates.

The way we use our buildings in the UK (our culture) is very different to that of the hotter middle Eastern or even Mediterranean countries. Elsewhere, buildings are shut up during the day to keep the heat out. Here, we like large windows and openings to allow light into our buildings and to provide views of the surrounding landscape, and we are more concerned about heat loss in winter than heat gains in summer. The introduction of double glazing, originally to provide noise reduction, is now standard due to its status as an energy saving feature. This, along with improved airtightness and increased insulation levels, are some of the reasons why we are now seeing instances of overheating in new buildings. Each pane of glass reduces the amount of light passing through a window by ~10% but glass being opaque to thermal infrared radiation stops ~50% of heat getting out again, a value which is often improved further through the coating of the glass with a thin layer of metal, so-called low-ε (emissivity) glass. This is why in the Nordic countries you will see triple and even quadruple glazing. You might have expected that the widespread adoption of double glazing would lead to a change in the way we design our buildings but this does not seem to be the case. Of course, this extra heat retention is a benefit in winter, but it can be a hindrance during warmer sunny weather, particularly when the Sun is lower in the sky (early morning or late afternoon, or during the shoulder months in spring and autumn) because solar radiation can penetrate further into a room, heating surfaces and reducing the likelihood that radiation (which is radiated in every direction) can get out of the building again. As the climate warms, design features such as double glazing, which are intended to keep heat in in winter, may become a major cause of overheating unless steps are also taken to keep heat out in summer. Shutters over windows, such as one sees in the warmer Mediterranean climates (see figures 4.18 and 4.20), would seem a logical solution; however, this would be a serious departure from the British culture. The challenge therefore is how to design new buildings and adapt existing ones to this changing climate without fundamentally changing the way we use our buildings.

There are several examples of successful attempts to produce climate change resilient buildings that combine aspects of both mitigation and adaption, both in domestic and non-domestic buildings. The recent 'Design for Future Climate' tranche of industrial research projects funded by the then Technology Strategy Board saw ~50 real building projects across the UK consider how to adapt to a changing climate. The projects ranged from eco-towns, to hospitals and laboratories, and from commercial offices to swimming pools. Sadly, although plans of adaptions to the building design had to be drawn up and timelines of any future interventions provided as a stipulation of the project funding, actual incorporation of these interventions into the constructed building was not required.

There are, however, several built examples of how to make buildings resilient to climate change, while at the same time reducing energy usage to satisfy the mitigation agenda. One such example, is Montgomery Primary School located in Exeter, which embodies many of the principles we have discussed throughout this book. As part of the UK government's 'Building Schools for the Future' incentive, several schools around Exeter and the Southwest were retrofitted and others demolished and rebuilt. However, this exercise was not overly successful, with several schools reporting overheating issues, requiring remedial works and leading to well publicised law suits between the local authority and the contractors. Montgomery School was designed in response to this, with the school designed to still pass current overheating criteria in the 2050s. The initial brief was also for the school not to have a heating system, with heat provided by the pupils and stored overnight and over the weekends. This was later relaxed but the school was designed to be zero-carbon in use. To achieve this, the school was designed according to the stringent Passivhaus principles and features triple glazing, a highly insulated and airtight building envelope and a mechanical ventilation system with heat recovery to extract heat from exhaust air to warm incoming fresh air. Montgomery Primary School was the first certified Passivhaus school in the UK, and has also been shown to be zero carbon in use (over the course of a year) thanks to the large array of photovoltaics (figure 8.1) on the roof and was awarded building of the year by CIBSE in 2014. While this is all very impressive, we are concerned here with its climate change resilience credentials.

Many Passivhaus buildings are lightweight in construction, not because the standard stipulates this but because the super-insulated façades tend to be wooden

Figure 8.1. View of the southern side of Montgomery Primary School in Exeter, showing the large array of photovoltaics and its shading effects. Photo taken midday 16 October 2012.

Figure 8.2. Cross sections of a typical super-insulated wall build up (left-hand panel) and the heavyweight walls used at Montgomery (right-hand panel). Both contain 150 mm (6′) of insulation.

frame filled with insulation (figure 8.2) to create very responsive buildings, which can be heated or cooled easily. Whereas in order to store metabolic heat from the pupils, we require a building with significant thermal mass. The entire school is made from concrete with 150 mm of expanded polystyrene insulation sandwiched between

150 mm (6′) of concrete on the inside and 75 mm (3′) of concrete on the outside surface (figure 8.2), the ceilings and floors are also concrete. The concrete is fair faced (smooth) and is not plastered, with only paint applied on the walls for aesthetic reasons and to improve the daylight uniformity in the classrooms.

All this concrete not only provides the building with a huge amount of thermal mass to regulate internal temperatures and to store metabolic heat, solar gains and other internal heat gains from lights and IT for example but it also has the benefit that cast concrete such as this is sufficiently airtight to meet the Passivhaus standard, while brick, block or wooden frame constructions such as that in figure 8.2 will require an additional airtight layer, either a membrane embedded in the construction or gypsum plasterwork on the interior surface. Concrete is also resistant to wind shear forces, driving rain penetration and flood damage. Montgomery Primary School is located close to the river Exe in an area that has historically been prone to flooding. For this reason, the new school building was raised 1 m above the level of the existing school; a tactic now employed in the Netherlands after the removal of dykes as part of the 'Room for the River' project.

Schools in the UK are some of the most stringently designed buildings with guidance and regulations pertaining to air temperatures, CO_2 concentrations, ventilation rates, acoustics and daylighting. This poses several problems; for example, in order to meet the daylighting requirements, windows have to be large, which if exposed to direct solar radiation can easily lead to overheating. Additionally, a standard 8×7 m classroom such as those at Montgomery, will contain ~30 pupils, at least one teacher, lighting and IT (e.g. PC and smartboard). This can easily equate to ~5 kW of heat being generated inside the room, which will need to be regulated by the structure or removed by the ventilation. It is for this reason that schools and offices are prone to sudden overheating when the Sun comes out. It was decided early on in the design process for all the classrooms to be located on the north façade, which is two storey, and offices and staff rooms to be located on the south façade. North facing windows in the northern hemisphere will only receive diffuse sunlight, so are ideal for daylighting purposes because there is limited risk of glare or overheating. Comparison of figure 8.3 shows that the windows on the southern façade are smaller and are shaded by wither brise-soleil or from the overhanging roof and photovoltaics. The photos shown here were taken during the official opening of the school on 16 October 2012, and hence the roof shading is not fully covering the south facing windows in the images but would do so in the late spring and summer months. This design of shading device allows passive solar heating of the building in the winter months when solar gains are beneficial but prevents direct solar radiation entering the building in the summer. Internal blinds (see figure 8.1) are provided to limit solar gains in the shoulder months if required and to reduce glare.

The activities within schools generate a large amount of heat, but they are also used primarily during the UK heating season. Therefore, controlling the flow of heat within the building is key not only to reducing heating bills but also to preventing overheating, both now and in the future. The high thermal mass structure of the building utilises fair faced concrete walls, a polished concrete floor (containing crushed glass to improve the aesthetics) and a concrete ceiling. All these hard

Figure 8.3. The northern facade of Montgomery school, showing the large windows for the classrooms.

parallel surfaces mean that measures need to be deployed to control the acoustics and reduce the reverberation time in order for the school to be a viable learning environment. This normally takes the form of carpets, soft furnishings and acoustic panels to absorb sound waves; however, these would also mask the thermal mass. This is often an issue with modern office buildings in that all the thermal mass is located within the floor slabs but is insulated from heat transfer by carpet on one side and ceiling tiles (to improve acoustics and hide services) on the other. In Montgomery Primary School, since walls are used for student work, acoustic panels had to be located on the ceiling; however, these took the form of rafts rather than a continuous ceiling to allow air circulation and heat transfer to the concrete above. This required some computer simulations, both to determine how much acoustic material had to be used to achieve the requires reverberation times in the classrooms and also the size, layout and separation of the individual rafts to allow the formation of convection currents within the classroom to allow heat from the pupils to be transferred into the floor/ceiling slab. The final result can be seen in figure 8.4, the rafts are able to hide the services and house the lighting fixtures while still allowing heat transfer between the air and all available thermal mass.

The majority of heat is generated in the classrooms; these are also the main occupied areas. We have already detailed what measures are taken to promote absorption of this heat into the concrete structure, but rather than waste heat that would otherwise be lost with exhaust air from the classrooms the school employs several other measures. The school, like many other Passivhaus', uses a mechanical ventilation system with heat recovery with windows only opened in summer to control overheating. While heat exchangers are efficient (~85%) at removing

Figure 8.4. Acoustic ceiling rafts in the sports hall suspended below the concrete ceiling to allow air circulation.

Figure 8.5. View from a corridor into a classroom during construction, showing holes for the 'vision panel' and for the air extract above.

sensible heat (and sometimes latent heat as well) from the exhaust air supply and transferring this to the incoming fresh air, it is more efficient to transfer as much heat to the structure as possible before air passes to the heat exchanger. As such, warm air is extracted from the back of the classrooms through openings visible in figure 8.5

Figure 8.6. Views along the corridors, ground floor on the left and upper floor on the right. Warm air is extracted from the classrooms and passes along the corridors and up through holes in the first-floor corridor.

into the corridors, which are unheated spaces. The warm air passes along the corridors, air from the ground floor is able to pass up through holes in the first-floor corridor (figure 8.6) before being extracted at high level. In this way, exhaust air has the maximum opportunity to give up its heat (both sensible and latent) to the thermal mass in the building structure before it reaches the heat exchanger. In order to prevent noise transmission between classrooms via the corridors, the ventilation holes are occluded with sound absorbing material, which can be seen in figure 8.6.

The holes in the first-floor corridor also allow daylight from rooflights above to penetrate into the back of the classrooms via 'vision panels', thereby improving daylighting uniformity in the classrooms. The ability to move heat around the building is particularly important, not just for reducing heating bills for the school but also in preventing overheating in the warmer months, both now and in the future. The openable rooflights allow heat to escape from the school via buoyancy driven stack ventilation, which does not require wind flow to be effective. The mechanical ventilation system allows the secure night-time purging of heat from the building without the opening of windows or rooflights. Removing the excess heat from the structure overnight refreshes the thermal mass, allowing it to store excess heat again the next day, reducing peak temperatures and preventing overheating. In this way the school can provide year-round thermal comfort, even in the height of a 2050s summer under an aggressive estimate of climate change.

Montgomery Primary School is a useful example to provide in the final chapter of this book because it utilises many of the principles we have covered in earlier chapters, albeit applied in way suitable for British architecture. The site utilises water permeable surfaces to reduce surface water flooding and is raised to prevent fluvial flooding. Gutters and downpipes are also oversized to cope with more intense rainfall in the future. The concrete structure is not only strong to resist strong winds but also resistant to driving rain and flood damage. The highly insulated and airtight shell keeps heat out in summer and keeps it in during the winter. Openings are shaded from direct solar radiation and the vast amount of thermal mass regulates internal temperatures, while the provision for fan night-time purging allows the building to remain comfortable, even during a prolonged heatwave event. While the building undoubtedly contains a huge amount of embodied carbon, it is highly efficient, is a net exporter of electricity to the national grid and does not overheat like its peers. The teaching staff have also noted an increase in student attainment since moving to the new school building, which can linked back to the improved internal environmental conditions.

Montgomery was a purpose designed new building but its successes show that it is possible to be both energy efficient and resilient to a changing climate. When it comes to our existing housing stock, there are several key principles, which will be discussed in the following subsections.

8.1.1 Shelter from the elements

Providing shelter is the fundamental purpose of our buildings. In order for this to continue, it may be necessary for us to provide shelter for our buildings from the extremes of weather. High winds and driving rain can cause damage to building fabric and possible structural failure. In exposed areas, it is a benefit to make buildings as aerodynamic as possible. Since wind speed increases with height, lower lying buildings such as bungalows will be more resistant to high winds than two storey or greater buildings. Thought should be given to the shape and orientation of the roof. For example, does the wind tend to come from a set direction? If so, then the area of the roof that faces directly into the wind should be reduced and an appropriate inclination should be chosen so that wind pressure is reduced. For existing buildings, wind breaks can be deployed, such as the planting of trees a suitable distance away or even the creation of grass banks upwind will disrupt wind flow and shelter the building from the high winds and driving rain.

Given the projected changes to rainfall intensity, the way we currently size our gutters based the amount of rainfall to fall in a minute, will mean that our current gutters even if kept clear will no longer be able to handle the volume of water running off our roofs. Overflowing gutters can lead to water running down the side of our buildings, which can lead to water seeping into our building and cause structural damage. Increasing the volume of the gutter or adding additional downpipes will help alleviate this problem. Alternatively, if the roof structure will take the additional weight, the use of vegetated roofs will slow rainfall runoff and allow existing gutters to cope with runoff (if any).

The increased intensity of rainfall may well require infrastructural upgrades to stormwater drainage. This can be alleviated at source (on-site) through the incorporation of rain gardens and more permeable surfaces rather than concrete and impermeable paving. This will reduce surface water runoff and help prevent flash flooding of properties. Elements of water storage such as ponds or water butts are also a good idea to cope with the drier summer months.

8.1.2 Keep heat out

It is much easier to keep heat out of our buildings rather than to try and remove it once its inside. Solar radiation can be up to 1000 W m^{-2} in the UK, which makes controlling the heat of the Sun the first line of defence when trying to prevent overheating, even for buildings with high internal gains such a schools and offices.

Reflecting solar radiation before it can be absorbed would seem to be the first step. Changing the emissivity of our buildings by simply painting them a lighter colour can help keep them cool. We only have to look at the many examples around the Mediterranean, such as Elba (figure 4.11), Cannes (figure 4.18) or Santorini (figure 4.21), to see how effective this can be.

The solar radiation that is not reflected is absorbed, raising the temperature of the absorbing surface according to its specific heat capacity, volume and conductivity. More insulative materials will attain higher surface temperatures and will radiate more heat to their surroundings with heat less able to pass through to the building interior. The most exposed part of any building to solar radiation is the roof, which is a large, unshaded and often optimally inclined space, receiving the maximum amount of solar radiation per m^2. Therefore, increasing the amount of insulation in a roof will help limit how much heat from absorbed solar radiation can pass into the spaces below. Vegetated roofs are also of benefit here. In addition, to their rainwater attenuation credentials, vegetation typically has a higher albedo than roofing materials, reflecting solar radiation back out to space and the evapotranspirative cooling effects will actively cool the soil and building fabric below.

Any openings in our buildings such as windows and French doors, need to be protected from direct solar radiation in the warmer months. East and west facing openings should be avoided if possible so that low angle solar radiation at the beginning and end of the day cannot penetrate deep into the building, where it is unlikely to be able to escape readily. At lower latitudes, south (or north) facing openings are less at risk because the solar altitude is so great; however, at mid-latitude locations such as the UK, shading is required to prevent intense solar radiation entering our buildings. Balconies, roof overhangs and fixed shades such as brise-soleil are the obvious choices to shade openings and façades, but can be difficult to implement in existing buildings and can be considered unsightly. An easy to implement alternative is the use of a veranda or trellis to support plants, which will shade the opening in summer but allow passive solar heating in winter. Plants such as clematis are ideal for this because they are light, good climbing plants which can be easily supported. The use of vegetation will also provide some air filtration and some evaporative cooling of incoming air.

Finally, we should consider the materials we surround our buildings with. Figures 4.13 and 4.14 provide some indication that hard impermeable surfaces around our buildings, particularly near openings, can lead to increased heating, even after the Sun has gone down. Ideally, patios or driveways should be a little way away from the walls so that radiated heat does not enter our buildings. There is also a noted difference in the temperatures of water permeable surfaces compared to impermeable ones, which combined with the benefits of reduced surface water runoff and increased percolation makes the choice of surface materials around out buildings fairly easy.

8.1.3 Remove excess heat

The removal of heat is key to providing thermal comfort and preventing summertime deaths in the future. Unfortunately, the highest temperatures and the most dangerous heatwave events are typified by a lack of wind to purge heat from our buildings. This means that even if we throw our windows open, we will struggle to eject the heat from our homes. Without resorting to a mechanical ventilation to purge the heat, we can instead utilise the density difference between air at different temperatures and heights. Buoyancy driven ventilation is often not utilised due to concerns over security—we do not like to leave windows open on the ground floor. However, the inclusion of secure openings, whether vents or louvered windows, will allow the flow of air through our buildings at night to remove heat (or even during the day when we are not there).

Increasing access to thermal mass within our homes can be beneficial in reducing peak temperatures. Replacing wooden laminate flooring and carpet with ceramic or stone flooring will increase the amount of available thermal mass and also allow for conduction of heat into the screed/concrete below. Perhaps while not desirable, the replacement of dry-lined partitions with brick or block ones with gypsum plasterwork will provide additional thermal mass to help regulate temperatures. For office buildings, simply removing ceiling tiles around the perimeter of an office or replacing them with egg crate ones will allow air to circulate in the void space above and transfer heat into the otherwise masked thermal mass of the floor/ceiling slab above.

8.2 Urban resilience

The ideas of urban resilience are principally an extension of the ideas for building resilience. The choice of materials across an urban area is a determining factor in how resilient it will be to climate change. Most cities are located along rivers, which are typically constrained as they pass through urban areas. Making an urban area more permeable to increase ground water percolation and slow and reduce surface water runoff will alleviate pressure on these rivers, reducing the likelihood that they will flood and cause widespread damage. In the Netherlands, the 'room for the river' project is moving or removing dykes to allow rivers room to flood, providing sacrificial areas for rivers to expand into. This is a similar principle to the flooding of the Somerset Levels and to a lesser extent the Exeter's flood relief channel, both mentioned in chapter 3, because they provide somewhere for excess water to go. Such measures, however, are not always possible, so instead we need to consider reducing the volume of water in rivers by other measures to limit fluvial flooding. Lessons can be learned from the destruction of Boscastle in 2004, where clearing of upstream scrub land meant that water entered the river faster, increasing the water flow rate trying to pass through the village. If rural areas upstream of urban areas can be hydrodynamically rough to slow water down, then water flow rates can be reduced and the likelihood of fluvial flooding reduced. Flash flooding from surface water is also a real issue in urban areas and is only likely to become more significant as rainfall intensities increase as a result of climate change. While permeable surfaces help reduce this and it is now in fact a policy of many local authorities to not permit the creation of impermeable driveways, for example, the water still needs somewhere to go if the ground becomes saturated or if storm drains are already full. The creation of areas designed to collect rainwater within the urban areas as part of a wider sustainable (urban) drainage system (SuDS), such as swales and rain gardens (whether grass or paved ones), will allow the retention of water for a short while so that river levels can drop and drains can empty.

If such measures are still insufficient, then it may well be necessary to build defences to keep the water at bay, such as raising the height of river banks and building walls. For some coastal villages such as Fowey in Cornwall, higher defences are the only real option to cope with rising sea levels and increasing storm surges. Whether these defences take the form of higher sea walls or tidal barrages such as those in the Thames or outside of Venice will depend on the individual circumstances of the urban area in question and available funds.

The urban heat island (UHI) is a product of the layout of the buildings in an urban area and the activities going on inside them. The increase in urban temperatures due to the heat island is equivalent to the most aggressive estimates of climate change over this century and the potency of the heat island is greatest during extreme heatwave events further increasing the risk to human life. To make matters worse, the formation of the heat island leads to the accumulation of airborne pollutants at street level. Therefore, the reduction of the UHI can be considered as a key urban resilience measure.

The two main ways to reduce the UHI are to remove heat from the urban area and to encourage vertical transport of air. Altering the albedo of our buildings and also roads and pavements will help reduce solar heating, encouraging the reflection of shortwave solar radiation back out to space before it is absorbed. The geometry of most urban areas leads to the absorption, radiation and reabsorption of longwave thermal radiation by roads and buildings. Allowing heat absorbing surfaces to shed this heat is key to reducing the UHI. Increasing the sky view factor of a street will aid in this. There is also much to be said for making cities greener and bluer. The Garden City movement devised by Ebenezer Howard in 1898 and realised in the cities of Letchworth and Welwyn Garden City was originally created as a way to overcoming the unhealthy conditions and overcrowding that existed in other cities at the time. This philosophy has now been adopted by other cities and towns, not just for the psychological and physiological benefits but also as a way to reduce the UHI. In 1967, the Prime Minster of Singapore announced the intention to turn the city-state into a garden city with abundant greenery and a clean environment to make life more pleasant for its people. This led to over 55 000 trees being planted in the next three years to 1970. In the mid-1970s, the creation of new parks became a major focus of this 'Garden City' vision. By 2014, the number of trees planted had increased to ~1.4 million and the number of parks had increased from 13 to 330, with an increase in the area of green space from 879 ha (~0.9 km^2) in 1975 to 9702 ha ($\sim$97 km^2) in 2014. The benefits of such widespread greening are not just associated with the evapotranspirative cooling of the atmosphere but also with an increase in surface roughness and perhaps more importantly an increase in the diversity of the roughness, leading to turbulence and increased vertical transport of air. Green space also typically has a higher albedo than other surfaces in urban area and the evapotranspiration of water is an endothermic process, both of which will reduce the amount of heat in an urban area. While the environmental physics behind the formation or disruption of a boundary layer over a city is complex and perhaps not fully understood, the incorporation of networks of green and blue space throughout an urban area will lead to an increase in vertical transport due to the fact that water vapour is less dense than air. Additionally, increasing turbulence through the incorporation of varied surface types, building shapes and sizes will lead to a greater diffusion of heat and pollutants into the atmosphere above.

Imagine a city that instead of being hard, grey and black, is soft, green and blue, with vegetation and water defining the form and character of the surrounding buildings. Such a city would be permeable to and absorbing of intense rainfall, and be cooled by the evapotranspiration of lush vegetation. Extensive green and blue spaces would provide valuable environmental capital, enhanced biodiversity and public amenities. The diverse urban surface would disrupt stable boundary layer formation, disperse air pollution and reduce urban temperatures to keep its citizens safe and comfortable at work, in their homes and on the streets. This, perhaps, is the city of the future!

www.ingramcontent.com/pod-product-compliance
Ingram Content Group UK Ltd.
Pitfield, Milton Keynes, MK11 3LW, UK
UKHW051855100926
8556IPUK00005B/640

9 780750 352